RESEARCH REPORT

Pathways from Climate Change to Conflict in U.S. Central Command

Nathan Chandler, Jeffrey Martini, Karen M. Sudkamp,
Maggie Habib, Benjamin J. Sacks, and Zohan Hasan Tariq

For more information on this publication, visit **www.rand.org/t/RRA2338-2**.

About RAND

The RAND Corporation is a research organization that develops solutions to public policy challenges to help make communities throughout the world safer and more secure, healthier and more prosperous. RAND is nonprofit, nonpartisan, and committed to the public interest. To learn more about RAND, visit www.rand.org.

Research Integrity

Our mission to help improve policy and decisionmaking through research and analysis is enabled through our core values of quality and objectivity and our unwavering commitment to the highest level of integrity and ethical behavior. To help ensure our research and analysis are rigorous, objective, and nonpartisan, we subject our research publications to a robust and exacting quality-assurance process; avoid both the appearance and reality of financial and other conflicts of interest through staff training, project screening, and a policy of mandatory disclosure; and pursue transparency in our research engagements through our commitment to the open publication of our research findings and recommendations, disclosure of the source of funding of published research, and policies to ensure intellectual independence. For more information, visit www.rand.org/about/principles.

RAND's publications do not necessarily reflect the opinions of its research clients and sponsors.

Published by the RAND Corporation, Santa Monica, Calif.

Library of Congress Cataloging-in-Publication Data is available for this publication.
ISBN: 978-1-9774-1242-3

ABOUT THIS REPORT

THIS REPORT PRESENTS an analysis of the relationship between climate change and conflict, and how that relationship is unfolding in the U.S. Central Command's (CENTCOM's) area of responsibility (AOR). First, this report analyzes what existing literature has identified as causal pathways from climate change to conflict. Second, the report details three illustrative case studies of climate-related conflict within CENTCOM.

This report is the second in a series stemming from a larger project to consider the impacts of climate change on the security environment in the region. The first report, *A Hotter and Drier Future Ahead: An Assessment of Climate Change in U.S. Central Command*, presents an analysis of projected climate impacts in the CENTCOM AOR in 2035, 2050, and 2070. The third report, *Conflict Projections in U.S. Central Command: Incorporating Climate Change,* generates ranged forecasts of future conflict in the region with climate change incorporated as one driver of that conflict. The fourth, *Mischief, Malevolence, or Indifference? How Competitors and Adversaries Could Exploit Climate-Related Conflict in the U.S. Central Command Area of Responsibility*, presents an analysis of how U.S. competitors—China, Russia, and Iran—may attempt to exploit climate-induced conflict in the CENTCOM AOR. And the final report, *Defense Planning Implications of Climate Change for U.S. Central Command*, analyzes "off-ramps" to climate-influenced conflict and the operations, activities, and investments CENTCOM needs to be prepared to execute, given climate impacts on the security environment. The primary audience for these reports is CENTCOM leadership, planners, and intelligence officers. The research reported here was completed in May 2023 and underwent security review with the sponsor and the Defense Office of Prepublication and Security Review before public release.

RAND National Security Research Division

This research was sponsored by CENTCOM and conducted within the International Security and Defense Policy Program of the RAND National Security Research Division (NSRD), which operates the National Defense Research Institute (NDRI), a federally funded research and development center sponsored by the Office of the Secretary of Defense, the Joint Staff, the Unified Combatant Commands, the Navy, the Marine Corps, the defense agencies, and the defense intelligence enterprise. For more information on the RAND International Security and Defense Policy Program, see www.rand.org/nsrd/isdp or contact the director (contact information is provided on the webpage).

Acknowledgments

This report benefited from the contributions of several colleagues from the RAND Corporation who are not listed as authors. Karen Edwards, Erik Mueller, Mark Toukan, and Steve Watts contributed to the literature review. Jessica Arana and Kristen Meadows assisted with design. Valerie Bilgri edited the report. And Rosa Maria Torres assisted with the formatting of this document. The authors also thank the RAND management teams that provided feedback on drafts of this report. Finally, we thank internal reviewer Bryan Rooney and external reviewer Abdulla Ibrahim, who provided insightful comments as part of RAND's quality assurance process.

CONTENTS

FIGURES AND TABLES

Figures

Tables

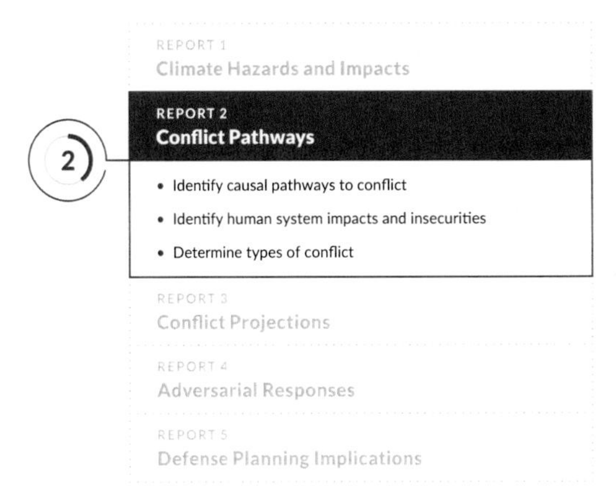

KEY FINDINGS

THE BROADER AIM of this project centers on characterizing the relationships between climate change and conflict to inform operational and longer-term decisionmaking by the U.S. Central Command (CENTCOM). This report, which builds on a semi-structured literature review and original case studies, analyzes the causal pathways from climate change to conflict.

- Although climate hazards may have direct impacts on violence, the pathways from climate events to war involve multistep processes in which the initial hazard typically triggers several intervening steps before manifesting as high-intensity conflict.
- The causal pathways from climate hazard to conflict vary but often begin with a hazard that results from a form of insecurity (such as food, livelihood, physical, or health insecurity), which then combines with climate impacts on state capacity, population flows, and other factors. When filtered through individuals' and armed groups' incentives to mobilize around greed or grievance, the combined impacts of these climate and non-climate factors culminate in conflict.
- The causal pathways from climate hazards to conflict below and above the thresholds of war are the same; what varies is the intensity of the ensuing conflict, not the path to get there.
- In total, our research identified seven broad families of causal pathways—and many more individual hypotheses—from which climate impacts could evolve into conflict.
- Climate-related conflict has already occurred in the CENTCOM area of responsibility, contributing to conflict below the threshold of interstate and intrastate war.
- Our research did not find a compelling case of past climate-related interstate war in the area of responsibility; however, there are plausible future contingencies for this outcome, based on our analysis of the defense acquisitions of potential disputants.

CHAPTER 1

INTRODUCTION TO CONFLICT AND CLIMATE CHANGE

AN ANALYSIS about how climate change may lead to conflict is presented in this report. Although climate-related conflict can occur anywhere in the world, our focus is on how this process has occurred and continues to evolve in the U.S. Central Command (CENTCOM) area of responsibility (AOR).[1] Much of CENTCOM is already coping with environmental stress caused by climate change and environmental management practices. Many of the factors associated with conflict (e.g., weak institutions, hybrid regimes) are present in the AOR, leaving the region vulnerable to the phenomenon of climate-related conflict.

In a prior report in this series, titled *A Hotter and Drier Future Ahead: An Assessment of Climate Change in U.S. Central Command*, RAND researchers found that the CENTCOM AOR will experience variable impacts from climate change depending on the subregion but that, in general, the physical environment in the AOR would become hotter and drier, with more exposure to extreme heat and drought.[2] In addition, sea level rise combined with land subsidence (i.e., land sinking from development and groundwater pumping) would lead to the inundation of low-lying areas, such as the Nile Delta and Al Faw Peninsula, where major rivers empty into the Mediterranean and Arabian Gulf, respectively. The same research also indicated likely increases in the incidence of extreme weather events in the AOR (e.g., extreme precipitation). See Figure 1.1 for the progression of the reports in this series.

This report begins by presenting an examination of what the academic literature identifies as causal pathways that lead from climate hazards to different types of conflict: intrastate conflict (also known as civil conflict) and interstate conflict. When possible, we further differentiate by the intensity of the fighting, using annual battle deaths of fewer than 1,000 as the threshold to differentiate the level of the conflict. Table 1.1 depicts the resulting typology, which differentiates conflict by primary actor and fighting intensity.[1] Although we find it useful to differentiate between these conflict types, the reader should note that conflicts can morph from one type into another (e.g., a conflict below the threshold of war may intensify until it becomes a full-fledged war or an intrastate war, such as a separatist struggle developing into an interstate war), blurring the distinctions between the categories.

Recent research has emphasized multistep causal pathways in which a climate hazard results in immediate impacts (e.g., crop failure, waterborne disease) that then pass through intermediate steps (e.g., loss of livelihoods, migration) before ultimately manifesting in conflict. Our research builds on this approach, examining a generalized, six-step process in which climate hazards interact with other variables to produce conflict. We present detailed causal maps in Chapter 2 for the intrastate and interstate domains, differentiating between the climate hazard that encompasses both stressors (e.g., year-over-year temperature rise) and shocks (e.g., monsoons), thereby

Table 1.1. Conflict Typology Used in This Report

	Intrastate Domain	Interstate Domain
Below the threshold of war	Riots, violent protests, limited campaigns of terror, proto-insurgencies, interpersonal violence, low-intensity intergroup violence	International crises, militarized interstate disputes
Above the threshold of war	Civil war, mature insurgencies, terror campaigns	Interstate war

NOTES: The examples for each category are not exhaustive. In the intrastate domain, some literature also differentiates by main conflict actors. Our review encompassed intergroup violence, interpersonal and intragroup violence, insurgencies, terrorism, and violent protests and riots.

Figure 1.1. Progression of Reports in This Series

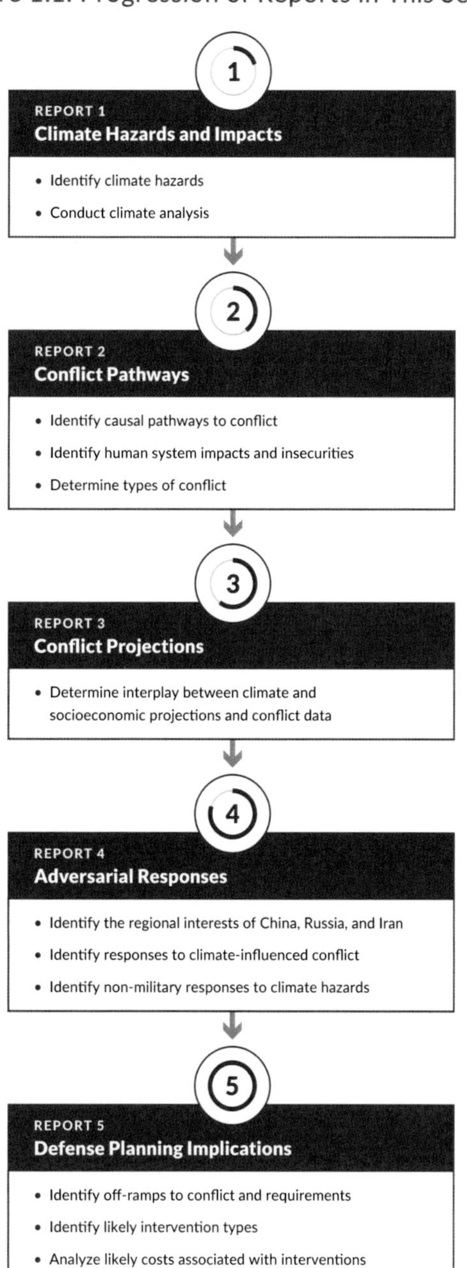

producing second- (e.g., crop failures) and third-order effects (e.g., food insecurity) before passing through what we call a critical pathway juncture, when hardships create a change in a conflict actor's capabilities or motivation to fight (e.g., increase in armed actor recruitment), which then interact with broader causal mechanisms (e.g., greed, grievances, security dilemmas) that culminate in a conflict. A simplified causal map illustrating this six-step process is depicted in Figure 1.2.

After identifying the causal pathways, our research explores the real-life ways in which environmental impacts contribute to conflict in CENTCOM. To do this, we profile examples of past or anticipated conflict in the region for the conflict types theorized above. The cases encompass a spate of violent protests in Basra, Iraq, that began in 2018 (intrastate conflict below the threshold of war), the 1971 civil war in former East Pakistan (intrastate war), and Egypt and Ethiopia's preparations for a potential future water war over the Grand Ethiopian Renaissance Dam (GERD) (interstate militarized dispute potentially escalating to interstate war). In none of these cases is climate change the sole cause for the resulting conflict; rather, the cases explore how environmental factors, which may be exacerbated by climate change, interact with other factors to contribute to conflict. This is consistent with the formulation that climate impacts can serve as a threat multiplier when other conditions make an area ripe for conflict.

The purpose of this research is to support CENTCOM command leadership, planners, and intelligence officers to prepare for a future security environment that is affected by climate change. Understanding the causal pathways from climate change to conflict should enable CENTCOM to anticipate how changes in the physical environment may reverberate in the security environment and when an area may be on a path to conflict or full-blown war that could lead to CENTCOM intervention. Readers should note that this portion of the research is not designed to estimate the *frequency* of such outcomes. Our estimates of the frequency of climate-related conflict will be presented in a forthcoming RAND report that uses machine learning to generate ranged forecasts for the incidence of conflict in the AOR.[3] Similarly, the implications for CENTCOM in terms of the operations, activities, and investments (OAIs) the command could take to mitigate the risk of climate-related conflict will be presented in our final report in the series.

Figure 1.2. Simplified Conceptualization of the Six-Step Process from Climate Hazard to Conflict

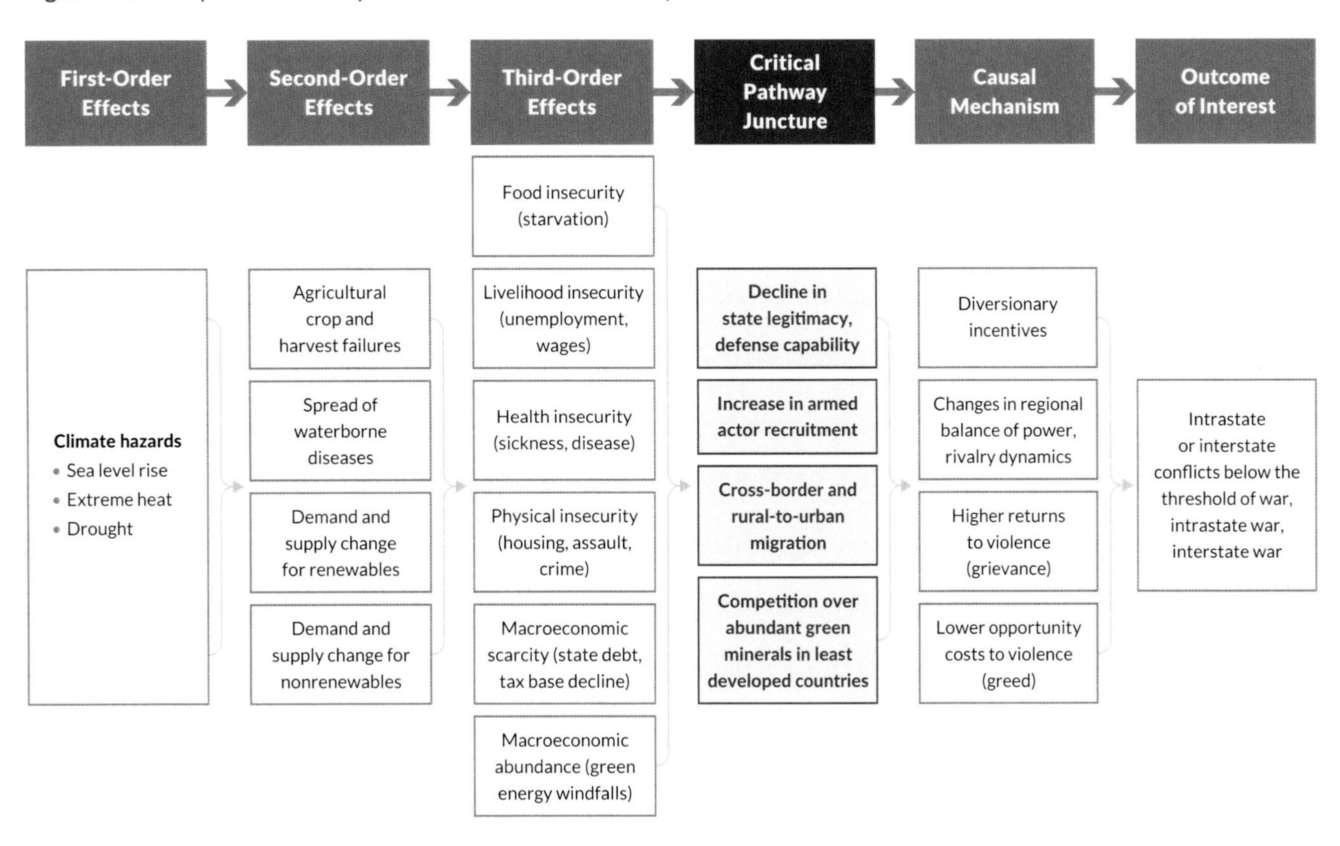

NOTE: This figure does not depict all the potential interactive effects across factors.

INTRODUCTION TO CONFLICT AND CLIMATE CHANGE

Endnotes

1 The CENTCOM AOR includes 21 states, including Egypt, the Levant, the Arabian Peninsula, Iran, and South and Central Asia. Israel was added to the combatant command in 2021 per changes in the unified command plan.

2 Michelle E. Miro, Flannery Dolan, Karen M. Sudkamp, Jeffrey Martini, Karishma V. Patel, and Carlos Calvo Hernandez, *A Hotter and Drier Future Ahead An Assessment of Climate Change in U.S. Central Command*, RAND Corporation, RR-A2338-1, 2023.

3 Mark Toukan, Stephen Watts, Emily Allendorf, Jeffrey Martini, Karen M. Sudkamp, Nathan Chandler, and Maggie Habib, *Conflict Projections in U.S. Central Command: Incorporating Climate Change*, RAND Corporation, RR-A2338-3, 2023.

CHAPTER 2

POTENTIAL CAUSAL PATHWAYS FROM CLIMATE CHANGE TO CONFLICT

HUNDREDS, IF NOT THOUSANDS, of scholarly journal articles have been written since the 1990s exploring the potential links between climate change–related hazards and the future of conflict. To review and distill the causal pathways contained in this vast body of academic literature, the RAND team adopted a three-step, semistructured approach. First, using bibliometrics as a guide, we selected seven of the most prominent and most influential review articles on the climate-conflict nexus published in the past ten years.[1] The original research in those seven review articles contained more than 400 unique sources that formed the basis of the sample of academic papers that we read for our literature review.[2] Second, since our research is focused on how climate hazards can result in different types of conflict, we categorized those articles based on the primary conflict type (intrastate or interstate conflict, above or below the threshold of war) they addressed.[3] Finally, we developed a framework for systematically coding the articles to identify the specific causal pathways posited in the literature.[4]

Main Branches in the Causal Pathway Literature

All causal pathways hypothesized and/or tested in the literature begin with either a climate change–related *stressor* (e.g., higher ambient temperatures, rising sea levels) or a climate change–related *shock* (e.g., flood, monsoon) as the first step in a causal chain. Then, the academic literature branches, as illustrated in Figure 2.1. One branch examines direct, one-step pathways, and a second branch examines indirect, multistep pathways.[5] Direct pathways are generally only posited as leading to individual-level conflict below the threshold of war (e.g., increased violent crime or interpersonal violence rather than increased intergroup conflict), so we suffice with summarizing those in the appendix. Indirect pathways can lead to full-blown intrastate or interstate war. The indirect branch can be further divided between scarcity-driven and abundance-driven multistep pathways. Because this branch is most relevant to CENTCOM, our report focuses on this portion of the literature.[6]

Figure 2.1. Conceptual Division of Climate-Conflict Literature

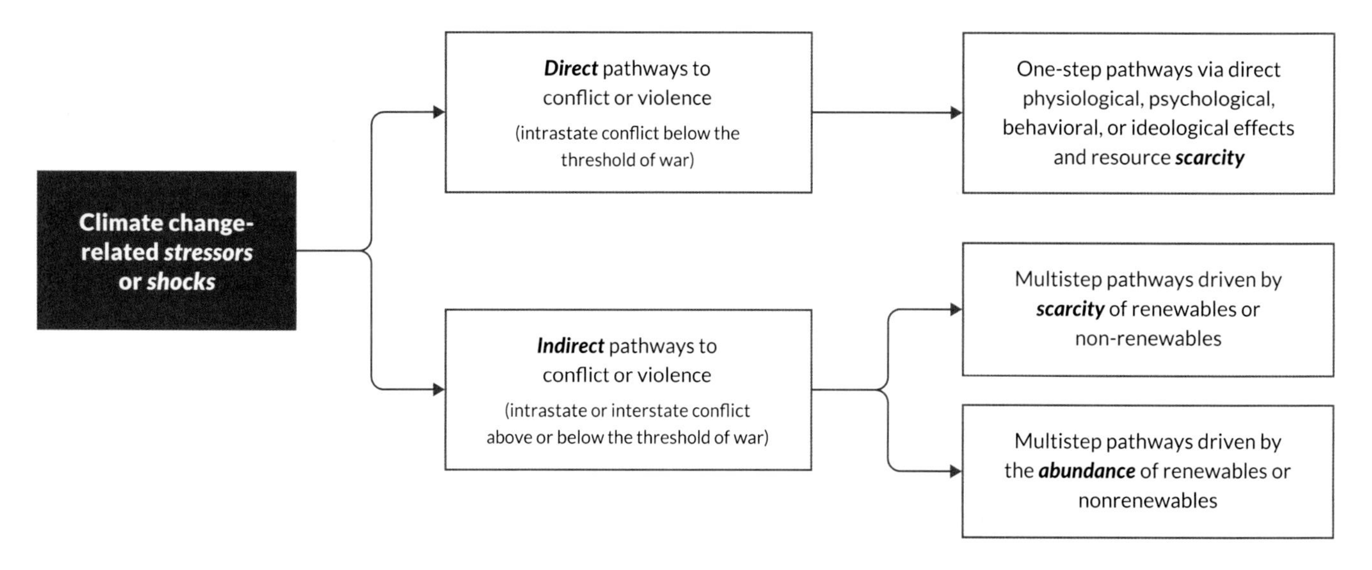

Indirect Pathways to Intrastate Conflict

We reviewed more than 200 peer-reviewed studies that, at least in part, traced indirect causal pathways from climate hazards to intrastate conflict below the threshold of war and/or to intrastate war. Our review focuses on the chain of events that follow a climate hazard; however, it should be noted that the one unifying conclusion drawn from these numerous studies is that climate change and the resultant impact on resources is "never a sole or sufficient cause of large migrations, poverty, or violence; it always joins with other economic political and social factors to produce its effects."[7] Climate change is generally seen as a threat multiplier rather than a direct cause of conflict.

In the sections that follow, we attempt to group related hypotheses into four broad families of multistep pathways that proceed from the effects of climate change: (1) increased intrastate resource curse dynamics via an abundance of renewable or nonrenewable natural resources, (2) increased intergroup and/or interpersonal competition for scarce resources, (3) enhanced violent nonstate actor (VNSA) capacity and/or capability, and (4) diminished state capacity and/or capability. Our review of the academic literature identified eight generalized hypotheses, which we have sorted across these categories as summarized in Table 2.1.

Increased Intrastate Resource Curse Dynamics via an Abundance of Renewable or Nonrenewable Natural Resources

Hypothesis 1a: *Climate change leads to increasing demand for renewable energy. Production of this energy creates a resource curse via large export windfalls of abundant nonrenewables (e.g., green minerals) that leads to intrastate conflict.*

We start with the branch in the empirical literature that postulates that climate change and conflict might be causally linked by *abundance* mechanisms rather than *scarcity* mechanisms. This school of thought posits that increased demand for renewable energy sources and technologies (e.g., electric vehicles, lithium batteries) will increase the likelihood of low-level violence, human rights abuse, and/or intrastate war via what has been called the *green energy resource curse.* The classic resource curse hypothesis—as advanced in the 1990s and the early-to-mid-2000s—argues that countries endowed with an abundance of nonrenewable resources (e.g., oil, gas, diamonds) are more prone to civil war and intrastate armed conflict because of at least three causal mechanisms:[8]

1. The discovery of and control over the resources is a prize that tends to become a proximate cause of intrastate war.
2. VNSAs tend to use illicit resources to self-finance their operations, and those resources prolong and/or intensify civil wars, insurgencies, and terrorist campaigns.
3. Government coffers tend to grow dependent on natural resource income streams while other sectors of the economy atrophy, rendering the state more vulnerable to commodity price fluctuations and increasing opportunities for government corruption, autocratic practices, and human rights abuses. In turn, weak institutions and poor governance capacity leave regimes more prone to violence.

More recently, this argument has been resuscitated by scholars who explore whether states that are endowed with an abundance of green mineral resources[9] may be increasingly vulnerable to resource curse–like pathways to intrastate conflict.[10] Church and Crawford explain this potential pathway as follows:

> As the demand for green energy technologies . . . continues to increase, so too does the demand for the minerals required to develop and deploy them. This growing demand should serve as an economic boon to those countries that are home to the principal reserves of strategic minerals for the transition, including cobalt, lithium, and rare earths. However, in countries struggling with political instability, where governance for the mining sector is weak, the extraction of these minerals could be linked to violence, conflict and human rights abuses.[11]

Like in traditional resource curse literature, the implication is that those states in the CENTCOM AOR with weak governance institutions, weak rule of law, and poor human rights records may be more likely to experience this pathway to conflict. But climate change–related resource curse dynamics resulting from abundant endowments of green minerals are not deterministic. Sociopolitical and economic characteristics matter, too. Barnett clarifies that in the abundance-driven paths to conflict, "The issue is not competition over scarce resources, but rather competition to gain dominant control over substantial income generating resources, or more equitable access to the spoils of resource extraction."[12]

***Hypothesis 1b:** The same as 1a, except this potential pathway operates through an abundance of renewables, such as hydroelectric power, rather than nonrenewables.*

The second abundance-driven hypothesis is the same as the first, except for one key difference: It focuses on lootable *renewable* natural resources rather than *nonrenewables*, just as the original resource curse literature evolved to include theories on pathways from an abundance of renewables (e.g., narcotics, timber, cash crops) to intrastate conflict. In the context of climate change, some works have posited that large-scale hydroelectric exports could produce similar resource curse effects as nonrenewable lootable resources.[13]

Table 2.1. Summary of Potential Indirect Pathways to Intrastate Conflict

Category	Hypothesized Pathway
Increased intrastate resource curse dynamics via an abundance of renewable or nonrenewable natural resources	**Hypothesis 1a:** Climate change leads to an increasing demand for renewable energy. Production of this energy creates a resource curse via large export windfalls of abundant *nonrenewables* (e.g., green mineral) that leads to intrastate conflict.
	Hypothesis 1b: The same as 1a, except this potential pathway operates through an abundance of *renewables*, such as hydroelectric power, rather than non-renewables.
Increased intergroup and/or interpersonal competition for scarce resources	**Hypothesis 2a:** Climate change-related scarcity of renewables or nonrenewables will increase food, livelihood, health, and/or physical insecurity. This will drive intrastate conflict because of increasing returns or decreasing opportunity costs to violence.
	Hypothesis 2b: Similar to 2a, except this pathway hypothesizes that intergroup conflict will be driven by the maldistribution of scarce resources among groups. These dynamics will fuel *grievance-based* intrastate conflict.
	Hypothesis 2c: Climate change-related scarcity leads to increased interpersonal gender and sexual violence *because of societal and/or cultural constructs*, fueling low-level civil violence and intrastate unrest via multistep social processes.[a]
Enhanced VNSA capacity and/or capability	**Hypothesis 3a:** Climate change-related scarcity leads to increased manpower recruitment in VNSAs via microeconomic mechanisms (e.g., wage, basic subsistence incentives), leading to more-capable VNSAs and more intrastate conflict.
	Hypothesis 3b: Same as 3a, except that increased manpower recruitment is driven by grievance-based mechanisms (e.g., radicalization, marginalization, migrant hardships) rather than economic incentives.
Diminished state capacity and/or capability	**Hypothesis 4:** Climate change-related scarcity will lead to state fiscal crises via the loss of revenue or new spending. Fiscal crises will lead to diminished state capability, popular grievances, and a decline in government legitimacy. These conditions increase the risk of intrastate conflict.

NOTE: Intrastate conflict encompasses both low-intensity conflict (protests, riots, etc.) and full-scale intrastate war.
[a] This hypothesis involves indirect, multistep pathways from climate change to increased interpersonal violence. It is distinct from the various hypothesized direct, one-step pathways from climate change to increased interpersonal violence discussed in the appendix.

Whether from an abundance of renewables or nonrenewables, pathways 1a and 1b both posit that new sources of wealth will catalyze resource curse dynamics, leading to increased intrastate conflict.[14] If increased demand for either renewables (such as hydroelectric power) or nonrenewables (such as green minerals) results in large-scale agricultural, mining, or industrial projects, the literature suggests that these negative resource curse effects likely would be localized in rural areas. This implies that any future intrastate conflicts that may develop in the CENTCOM AOR because of resource curse dynamics may manifest as protests or violence between national elites in the "center" and poor rural dwellers on the "periphery."[15]

Increased Intergroup and/or Interpersonal Competition for Scarce Resources

Hypothesis 2a: *Climate change–related scarcity of renewables or nonrenewables will increase food, livelihood, health, and/or physical insecurity. This will drive intrastate conflict because of increasing returns on or decreasing opportunity costs to violence.*

The origins of the environmental scarcity conflict literature can be traced to Malthus's canonical—albeit misguided—theory, which posited that exponential human population growth would test the limits of linear food production growth and lead to global food scarcities.[16] The Malthusian thesis predicts that mass deaths will occur from war, famine, or disease or that societies will adopt curbs on population growth.[17] During the "environmental awakening" of the late 1960s and early 1970s, a neo-Malthusian strain was resurrected in the empirical literature "predicting that the rapidly growing world population would soon exceed the resource base and lead to serious environmental destruction, widespread hunger, and violent conflicts."[18] Urdal further explains that "according to the neo-Malthusian conflict scenario, population pressure on natural renewable resources makes societies more prone to low-intensity civil war."[19] This second generation of scholarship extends Malthus's logic beyond the scarcity of renewable resources (food) to include the depletion of nonrenewables, such as coal, oil, gold, iron, and minerals (i.e., demand-induced scarcity).[20]

With broader acceptance of the human causes of global warming emerging in the 1990s, the neo-Malthusian school of thinking evolved further, explicitly linking climate change to the possibility of increased future conflict via demographic pressures on the environment. The RAND team reviewed roughly 100 articles examining how such scarcity-related mechanisms may lead to conflict below the threshold of war. The pathway was prominently advanced by Homer-Dixon and colleagues' seminal (albeit extensively challenged[21]) theoretical framework and casework.[22] This school of thinkers posits that human activities and demographic pressures will lead to resource depletion and the scarcity of renewables (particularly resources associated with food production, such as cropland, freshwater, forests, and fisheries), which will lead to internal armed conflict, including contributing to insurrections, ethnic clashes, urban unrest, violent protests, coups, assassinations, insurgencies, and other forms of civil strife or violence, especially in the developing world. According to Homer-Dixon's concept of "resource capture," climate change–induced scarcity is likely to encourage "powerful groups to capture environmental resources and [prompt] marginal groups to migrate to ecologically sensitive areas. These two processes in turn reinforce environmental scarcity and raise potential for social instability."[23] Extending Homer-Dixon's logic, Koubi explains that

> [c]limate change affects the likelihood of intragroup violence via the scarcity of renewable resources, such as freshwater, arable land, forests, and fisheries. Following a neo-Malthusian line of argument, it is assumed that adverse climatic conditions, e.g., high temperatures or low rainfall, coupled with overpopulation, reduce the resources needed to sustain human livelihood. Reduced resources increase competition, which leads to conflict (Homer-Dixon 2001). At the national level, for instance, less rainfall or high temperatures could lead to conflict among consumers of water, e.g., farmers and herders, as well as urban unrest, insurrections, and other forms of civil violence, especially in the developing world.[24]

One rich strain in this literature examines the risk of intergroup clashes, raids, or pastoral violence over access to increasingly scarce grazing and arable land, water holes, livestock, or cropland.[25] Bächler explains that "heavy environmental degradation can be a source of ethnic tensions, when ethnic groups with different socioeconomical traditions share a sensitive ecoregion."[26] Meier, Bond, and Bond argue that these pressures could produce dynamics by which stronger groups prey on weaker groups to capture resources during periods of scarcity.[27] As Vestby notes, the most commonly cited determinative link in these causal chains is that scarcity-induced hardships (e.g., food, livelihood, health, physical insecurity) effectively either decrease the opportunity costs of engaging in violence or increase the returns on violence.[28]

A second major variant on this hypothesis in the literature posits that food price shocks and/or food scarcity will fuel civil unrest or political violence in urban areas (but not rural ones). According to this line of argument—as tested by Smith—rural populations are generally food producers and thus are better suited to weather price shocks.[29] In contrast, future food price shocks that are the result of climate change are hypothesized to have a more deleterious effect in urban areas, where the population is generally composed of food consumers. As a result, this school of thinkers predicts that climate change–related price shocks will increasingly fuel civil unrest (protests, riots, etc.) in urban areas that could potentially devolve to low-level violence and increased domestic political instability.

Finally, a notable feature of the neo-Malthusian water scarcity literature is that pathways to *intrastate* water wars are generally viewed as less likely than pathways to *interstate* water conflicts (discussed in the next section). At local levels, some evidence suggests that increased cooperation may be more likely than conflict.[30] However, as Scheffran and Battaglini posit, cross-border migration could lead to secessionist insurgencies via water scarcity mechanisms:

> In this view [of water scarcity as a form of wealth deprivation], uneven water distribution may induce migration or the quest for resources in neighbouring regions. A disadvantaged group could seek to displace another group from a more water-rich territory or a water-rich region could secede or otherwise distance itself from central government control.[31]

Hypothesis 2b: *Similar to 2a, except this pathway hypothesizes that intergroup conflict will be driven by the maldistribution of scarce resources among groups. These dynamics will fuel grievance-based intrastate conflict.*

Some scholars posit that ecological marginalization of ethnic groups and the maldistribution of resources between the haves and have-nots of society will lead to scarcity-driven intergroup violence, similar to the concept of resource capture presented in hypothesis 2a. However, whereas the previous scarcity-driven pathway relates to predation and postulates that increased civil conflict will be fueled by microeconomic risk and reward mechanisms (i.e., increasing the returns on or decreasing the opportunity costs of intergroup violence), hypothesis 2b posits that increased intergroup civil conflict will be fueled by *grievance-based* dynamics. The RAND team reviewed dozens of articles scrutinizing the pathways that lead from climate change–related scarcity to grievance-fueled or justice-fueled political unrest or violence. According to the causal logic of this hypothesis, as climate change causes agricultural losses, environmental disasters, and/or the loss of habitable ecosystems, food, livelihood, health, and physical insecurity will increase in affected parts of the world. This theory predicts that broader popular scrutiny and anger over the inequitable distribution of resources in society will contribute to increasing political violence. These dynamics are hypothesized in the literature as potentially leading both to conflict below the threshold of war and to full-blown intrastate war. As Yeeles explains, another causal mechanism is "non-universal redistribution, inequalities, and ultimately grievances," from which climate change–induced scarcity and hardships may lead to the violent appropriation of resources.[32]

As a general proposition, the literature assumes that climate-driven scarcities will not affect everyone equally. Gleditsch points out that the effect will likely be felt differently between urban and rural communities; between local, regional, national, and global stakeholders; and between distinct ethnic, economic, political, and social haves and have-nots.[33] Fjelde and von Uexkull explain that the risk of communal conflict may be "amplified in regions inhabited by politically excluded ethno-political groups."[34] More broadly, marginalized ethnic, social, political, and economic groups might be excluded from the distribution of limited state or private resources that are made available to mitigate climate change–induced hardships, potentially fueling intergroup grievances.

In an examination of the Darfur civil war, Selby and Hoffman add nuance to this argument, observing that conflict dynamics are driven by corresponding perceptions of scarcity in one region and abundance in another. These perceptions drive those living in a region where resources are scarce to fight those living in areas that are more resource rich.[35] Ide et al. posit that extreme events (such as natural disasters) can lead to the exclusion of some ethnic groups, thereby fueling intrastate conflict; because armed conflict also raises vulnerability to natural disasters, this can become a vicious, self-reinforcing dynamic.[36] Ide et al. write that

> [g]rievances are predominantly linked to the perceptions of socio-economic and/or political injustices as causes of armed conflict. Several mechanisms can be involved in the creation or intensification of grievances after climate-related disasters. Prominent among them are perceptions of unequal distribution of disaster-related vulnerability, deprivation, relief, or reconstruction support. Disaster impacts often reflect if not intensify pre-existing inequalities and make them more acute and/or visible due to the disaster's magnitude.[37]

Within this hypothesis, we reviewed at least 20 articles with the hypothesis that migration will contribute to grievances that will lead to competition over resources and more identity conflict.[38] For example, existing groups might target new arrivals as retribution for straining state capacity and to deter future migrant flows into their communities. As Koubi explains,

> The influx of large numbers of 'environmental migrants' is likely to burden economic and resource bases in the receiving areas, thus promoting contests over scarce resources. . . . In addition, environmental migration could lead to conflict by stirring ethnic tensions that arise when migrants and residents belong to different ethnocultural groups and the arrival of newcomers upsets an unstable ethnopolitical balance.[39]

Ide et al. explain that the climate-related trigger for such ecomigration is often a natural disaster.[40] But mass migration might also be the product of slower-moving climate processes and environmental change. Some prominent research has focused on the role that this dynamic may have played in increasing violence after the start of the Darfur civil war in the mid-2000s. In the words of McDonald, proponents of this pathway "attempted to link conflict in Darfur to the manifestations of climate change, with climate-induced agricultural challenges encouraging population movements which were seen as triggering confrontation between groups over increasingly scarce natural resources."[41] As Bogale and Korf argue, "environmental scarcity when it coincides with socio-economic processes of rent-seeking and exclusion triggers political conditions ripe for violent struggles."[42]

Hypothesis 2c: *Climate change–related scarcity leads to increased interpersonal gender and sexual violence because of societal and/or cultural constructs, fueling low-level civil violence and intrastate unrest via multistep social processes.*

Another strain in the growing body of empirical research on climate-induced interpersonal violence focuses on the pathways by which climate change may lead to increased domestic and sexual violence directed at women and girls as a specific out-group. This topic may be particularly germane to future political stability in the CENTCOM AOR, given the theater's many conservative religious and cultural communities. Fröhlich and Gioli theorize that

> [w]ith increasing scarcity and degradation of land and water, those who are poor in resources, income, and power—many of them women—lose their rights to use these existential resources. The loss of livelihood due to environmental change, regardless of whether it was caused mainly by global warming or more by bad governance, is often the starting point of resource-related conflicts. . . . Such escalation processes have gender-differentiated causes and consequences: each societal group is affected differently both by environmental change and by conflict, depending on its specific position in the respective structures along which access to resources, income, and decision-making power is distributed.[43]

Nguyen similarly hypothesizes that existing vulnerabilities and inequalities that are "inherent in the social construction of gender prior to [a climate change–related natural disaster or extreme event]" will become "sharpened as efforts to survive become more urgent [in the wake of the event]," leading to heightened patterns of interpersonal violence against women and girls.[44]

Enhanced Violent Nonstate Actor Capacity and/or Capability

Hypothesis 3a: *Climate change–related scarcity leads to increased manpower recruitment in VNSAs via microeconomic mechanisms (e.g., wages, basic subsistence incentives), leading to more-capable VNSAs and more intrastate conflict.*

Terrorism experts began warning in the early-to-mid-2000s that climate change–related scarcity dynamics may engender feedback loops, by which "[c]limate change will provide the conditions that will extend the war on terror. . . . You have very real changes in natural systems that are most likely to happen in regions that are already fertile ground for extremism."[45] The essential progression of this pathway begins with some climate change–related shock or stressor (drought or persistent arid temperatures above the mean) that leads to microeconomic or macroeconomic secondary effects (such as agricultural crop failures, lower wages, unemployment, and food price spikes), which in turn produce perceived hardships on local populations (e.g., food insecurity, livelihood insecurity, health insecurity, physical insecurity).[46] At this critical juncture in the causal chain, the pathway posits that individuals are likely to join armed groups for simple livelihood and survival reasons—as opposed to purer ideological reasons—leading to an increase in armed group activity. According to this line of thinking, climate change–related shocks might lower the relative returns on agricultural labor (i.e., wages) or increase livelihood insecurity, which in turn will increase the propensity of individuals to join insurgent or terrorist groups in developing countries.[47]

Hypothesis 3b: *The same as 3a, except that increased manpower recruitment is driven by grievance-based mechanisms (e.g., radicalization, marginalization, migrant hardships) rather than economic incentives.*

Closely related to the previous pathway, another variant poses that similar conditions could lead to increased recruitment and manpower levels for VNSAs via grievance-based mechanisms. According to this formulation—as Bakaki and Haer and Heslin posit—climate change–related scarcity will create the unequal distribution of limited resources, either via government or via unofficial channels. The haves will fare better than the have-nots, with some out-groups harboring intensifying grievances. This school of thinking argues that such sources of ecomarginalization could radicalize more individuals and drive them to join VNSAs or armed political groups on the basis of grievance-based mechanisms (rather than simply out of desperation for a paycheck or subsistence). Heslin (2021) argues that climate change–related food price spikes—whether caused by local or international intermediatory steps—provide opportunities for opposition party leaders to channel related grievances into recruitment and mobilization, thereby increasing the power of opposition political movements and increasing the probability of civil unrest or low-level violence.[48]

A particular strand of this work focuses on how far-left terrorist groups may be able to relate climate change–based grievances back to their original causes (e.g., land reform, egalitarianism) and thus bolster recruitment. Kingdon and Gray profile three far-left movements that use political violence—the Revolutionary Armed Forces of Colombia (FARC), the Shining Path in Peru, and Naxalites in India—to show how climate change has furthered each group's recruitment.[49] The primary cause of these groups is not combating climate change; rather, the causal path is based on climate hazards exacerbating grievances that align with these movements' overarching frames and thus expanding the pool of potential recruits.

We note that another form of grievance treated in the literature is the hardship endured by climate migrants. Migration can foster ethnic tensions in receiving areas "as a result of the migrant being viewed as the 'other,'" leading to radicalization and conflict.[50] As Burrows and Kinney explain,

> "Otherness" may also go beyond the individual level and extend to issues of national identity, particularly if the migration is across international borders. Receiving countries may feel overwhelmed and potentially threatened by the influx of people with different languages and religions. More broadly, it is significant to note that uncertainty about the future is one of the most crucial factors that can lead to violent conflict, and in some ways perceived insecurity is more critical than actual insecurity. As such, even if in reality migrants do not pose a significant threat to political or economic power, the perceived risk may be enough to provoke conflict.[51]

Reuveny explains that "rebels may mobilize poor and frustrated rural migrants to challenge the state, which may respond with force."[52] Smith points out that climate change–induced mass migration will also create lucrative opportunities for human trafficking networks; this, in turn, could provide funding streams for VNSAs to strengthen their organizations.[53] Rural-to-urban migration can also enable VNSA recruitment, although Reuveny contends that "urban settings may offer migrants more opportunities, defusing tensions."[54] As argued by Schon and Nemeth, climate change increases the risk of terrorism via rural-urban migration. This internal migration is poised to accelerate urbanization, and rapid population influxes are difficult for cities to manage effectively. As a result, slums and urban ghettos have formed, are likely to continue forming, and are likely to grow where they already exist. These spaces often contain dense, opaque networks that states struggle to monitor. These networks are heavily populated by people who are disadvantaged or frustrated by their perceived exclusion from economic and political power. In these social environments, terrorist groups have many options for recruitment.[55]

Diminished State Capacity and/or Capability

Hypothesis 4: *Climate change–related scarcity will lead to state fiscal crises via loss of revenue or new spending. Fiscal crises will lead to diminished state capability, popular grievances, and decline in government legitimacy. These conditions increase the risk of intrastate conflict.*

Climate change–related disruption to agricultural output or other major economic activities could drive government balance sheet shortfalls because of falling tax revenues, placing strains on public spending and undermining the government's ability to provide services, including the effective provision of security. In turn, a weakened state is presumed to be an easier target for regime change or other violent insurrection. Kahl postulates, that in some cases, failed states will result.[56]

As Homer-Dixon observed in 1994, "severe environmental scarcity" could "simultaneously increase economic deprivation and disrupt key social institutions, which in turn would cause 'deprivation' conflicts such as civil strife and insurgency."[57] More than a decade later, Homer-Dixon quoted a panel of 11 retired generals and admirals warning that

> climate change is already a "threat multiplier" in the world's fragile regions, "exacerbating conditions that lead to failed states—the breeding grounds for extremism and terrorism." . . . Climate change will help produce the kind of military challenges that are difficult for today's conventional forces to handle: insurgencies, genocide, guerrilla attacks, gang warfare, and global terrorism.[58]

Research such as Buhaug et al. has analyzed whether the weakening of government via livelihood losses or other macroeconomic mechanisms (e.g., reduced tax base, commodity price crashes) could fuel grievances for conflict or unrest, particularly if scarce government resources and assistance are not distributed equitably among the population.[59] Dupont and Pearman postulate in part that climate stressors and shocks will result in "adding to the burden on poorer countries and stretching the resources and coping ability of even the most developed nations"[60] Barnett and Adger investigate whether climate change will increasingly harm state capacity to deliver the public goods and services that support their citizens and "help to maintain and build peace," thus risking higher rates of violent conflict.[61] Moreover, the weakening of government capacity may contribute to a negative feedback loop, limiting the state's ability to mitigate climate change–related hardship through accommodation and innovation.

Numerous studies have also hypothesized that increases in emergency government spending (rather than decreases in income) because of climate change–related environmental disasters, mass migration flows, or other macroeconomic shocks will create fiscal debt crises, particularly in less developed countries. According to this line of thinking, as climate-related migrant inflows intensify and overwhelm the capacity of state services, ethnic preferences or elite corruption might fuel the maldistribution of scarce resources, fanning grievances.[62] These impacts may change individual incentives for participating in unrest or violent action and ultimately change the power balance between opposition factions, armed groups, and the government, which can lead to the onset of conflict or the intensification of existing conflict.[63]

For this pathway, the literature also suggests that the political-economic impacts on the state from climate change–related natural disasters may function as "critical junctures" (tipping points) to conflict or may create "path dependent" trajectories to conflict.[64] Pelling and Dill explain that one school of thought "sees disasters producing an 'accelerated status quo'—change is path dependent and limited to a concentration or speeding up of pre-disaster trajectories which remain under the control of powerful elites both before and after an event"; they also explain that a second school "sees evidence that disasters can catalyse a 'critical juncture'—an irreversible change in the direction or composition of a political regime (or its subsets)."[65] The first dynamic (path dependency and accelerated status quo) theoretically functions to *concentrate* "established political and associated economic and cultural power" after a disaster, while the second dynamic (critical junctures and tipping points) theoretically functions to *contest* such existing power networks.[66]

Eastin similarly argues that natural disasters—which can occur as a result of climate shocks—may exacerbate existing conflict by harming state capacity to combat armed opposition while improving the ability of armed groups to evade defeat and capture.[67] Eastin posits that natural disasters harm the state's ability to finance counterinsurgency activities by forcing the state to spend funds on relief and reconstruction instead of on armed campaigns and to redirect troops from fighting to relief efforts, in addition to the damage caused to infrastructure that harms the state's ability to complete missions. Indeed, a debate among scholars has emerged about whether climate change–related drought in Syria and Iraq fueled Syria's civil war, which began in 2011. The drought drove the collapse of Syria's agricultural sector, resulting in widespread livelihood and food insecurity and mass migration, exacerbating already poor governance institutions and further weakening state capabilities.[68]

Figure 2.2 summarizes the causal pathways for both conflict below the threshold of war and intrastate war.

Indirect Pathways to Interstate Conflict

In contrast to the rich academic literature that treats the relationship of climate change to conflict as below the threshold of war and intrastate war, there is relatively scant original research that explores the intersection of climate change with interstate war or militarized interstate disputes as below the threshold of war. Among our sample of several hundred

Figure 2.2. Causal Map of Indirect Pathways to Intrastate Conflict

First-Order Effects → **Second-Order Effects (Interactive)** → **Third-Order Effects** → **Critical Pathway Juncture** → **Causal Mechanism** → **Outcome of Interest**

First-Order Effects

Stressors
- Mean ambient heat and cold
- Precipitation variability, aridity
- Sea ice retreat
- Sea level rise
- Ozone depletion
- Erosion, gradual land and ecosystem degradation

Shocks
- Natural disasters and extreme events
- Drought
- Inland flooding
- Coastal flooding
- Storms (tropical)
- Storms (winter and ice)
- Wildfires
- High-wind events
- Ocean current and weather anomalies

Second-Order Effects (Interactive)
- Loss of potable freshwater
- Agricultural crop and harvest failures
- Spread of waterborne diseases
- Livestock and fisheries sickness and death
- Loss of habitat, ecosystem, biodiversity
- Destruction of shelter, property, infrastructure
- Commodity price shocks
- Labor demand and wage shocks
- Demand and supply change for nonrenewables
- Demand and supply change for renewables

Third-Order Effects
- Food insecurity (starvation)
- Livelihood insecurity (unemployment, wages)
- Health insecurity (sickness, disease)
- Physical insecurity (housing, assault, crime)
- Macroeconomic scarcity (state debt, tax base decline)
- Macroeconomic abundance (green energy windfalls)

Critical Pathway Juncture
- **State fiscal or political crisis, decline in legitimacy and capacity**
- **VNSA recruitment, capacity-building**
- **Competition for scarce resources (resource capture)**
- **Maldistribution of resources, relative deprivation and marginalization**
- **Cross-border and rural-to-urban migration**

Causal Mechanism
- Higher returns to violence (grievance)
- Lower opportunity costs to violence (greed)

Outcome of Interest
- Civil conflict below the threshold of war or intrastate war

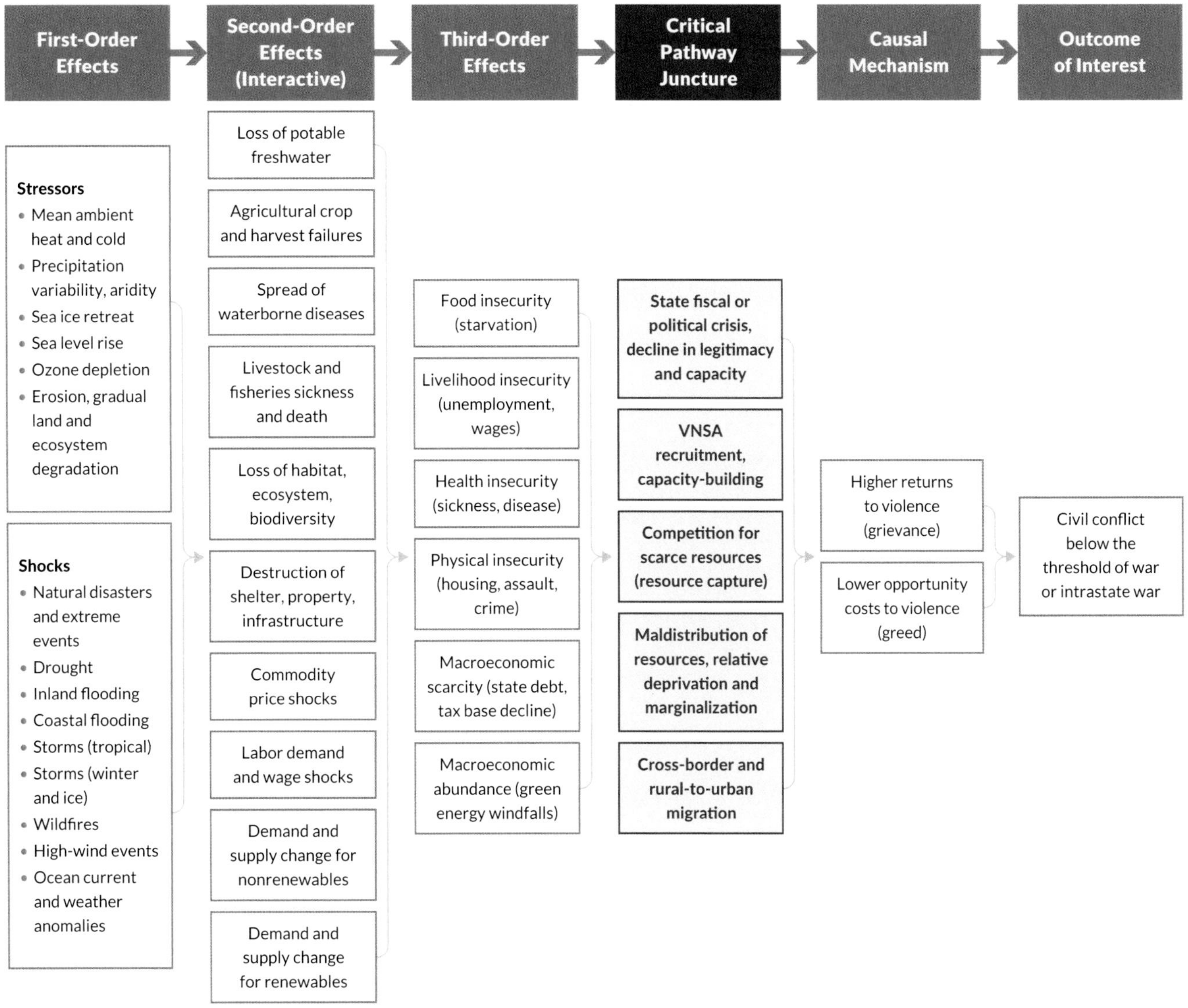

NOTE: This figure does not depict all the potential interactive effects across factors.

articles, only about 40 explored explicit pathways to interstate war.[69] Many prominent theorists assert that pathways to interstate conflict are less likely than pathways to intrastate conflict.[70] We identified eight pathways linking climate change to potential interstate war—or at least linking interstate crises with the potential to escalate to war—which we have categorized into three broad families of pathways: (1) increased resource curse dynamics via an abundance of renewables or nonrenewables, (2) increased interstate competition for scarce resources, and (3) increased risk from security dilemmas, diversionary incentives, and/or inadvertent escalation. We summarize these pathways in Table 2.2 and detail them in the remainder of this chapter.[71]

Increased Interstate Resource Curse Dynamics via an Abundance of Renewables or Nonrenewables

Hypothesis 5a: *Climate change will greatly increase demand for renewables. Hydropower development in transboundary basins will lead to conflict over access rights and escalate to interstate war over hydropower authorities and resource control (i.e., internationalized resource curse).*

The classic resource curse literature mainly explores relationships between an abundance of lootable and/or extractable resources (whether renewable, such as narcotics, or nonrenewable, such as fossil fuels) and *intrastate* war. A sliver of the literature extends those same dynamics to how they might

Table 2.2. Summary of Potential Indirect Pathways to Interstate Conflict

Category	Hypothesized Pathway
Increased interstate resource curse dynamics via an abundance of renewable or nonrenewable natural resources	**Hypothesis 5a:** Climate change will greatly increase demand for renewables. Hydropower development in transboundary basins will lead to conflict over access rights and escalate to interstate war over hydropower authorities and resource control (i.e., internationalized resource curse).
	Hypothesis 5b: Climate change-related competition over the control of abundant nonrenewable green minerals and/or the formation of strategic mineral cartels will drive interstate conflict that escalates into war.
Increased interstate competition for scarce resources	**Hypothesis 6a:** Climate change-induced scarcity of renewable natural resources will provoke interstate simple-scarcity conflicts, or zero-sum resource wars. Specifically, climate change-related scarcity of shared fresh water will increase tensions between nations straddling transboundary basins and lead to water wars.
	Hypothesis 6b: Climate change-related cross-border migration will exacerbate interstate rivalries via scarcity or security mechanisms, potentially escalating to grievance-based interstate war.
	Hypothesis 6c: Climate change-related shifts in natural boundaries could lead to conflict that escalates into interstate war over territorial disputes.
Increased risk from security dilemmas, diversionary incentives, and inadvertent escalation	**Hypothesis 7a:** The climate change-related weakening of state capacity could shift the underlying balance of power, leading to interstate war.
	Hypothesis 7b: Climate change-related extreme weather events could incentivize diversionary wars.
	Hypothesis 7c: The climate change-related search for energy independence leads to the proliferation of civil nuclear programs and the increased risk of nuclear incidents that inadvertently escalate to interstate war.

NOTE: Interstate conflict encompasses both low-intensity interstate crises and militarized interstate disputes, as well as full-scale interstate war.

apply to *interstate* war.[72] A limited amount of climate-conflict research has probed the hydropower and security hypothesis. According to this line of thinking, interstate water wars could result not from *scarcity*-driven mechanisms but from *abundance*-driven (macro) hydro-political mechanisms. Thus, the development and management of profitable, renewable hydropower—the world's largest source of renewable electricity—could lead to disputes over water rights, bargaining failures, and interstate crises or conflicts. As Sovacool and Walter point out, "Hydropower dams, the thinking goes, are one of the critical factors that can tribute to [more internal and external conflict in hydropower states]. . . . Other literature discusses how dams can become focal points during international disputes or targets for terrorism."[73] Hensel et al. adopt a similar logic, explaining that the "presence of hydroelectric projects on the river will increase the value of the water supply to one or both sides."[74]

***Hypothesis 5b:** Climate change–related competition over the control of abundant nonrenewable green minerals and/or the formation of strategic mineral cartels will drive interstate conflict that escalates into war.*

Another school of thought posits that the increased demand for renewable or clean energy sources and technologies (e.g., electric vehicles, lithium batteries) will increasingly drive interstate competition for access to nonrenewable green mineral resources, rendering national economies more susceptible to mineral price shocks, raising international tensions, and potentially leading to escalatory interstate conflict.[75] O'Sullivan, Overland, and Sandalow have theorized that the 21st century might see the birth of new strategic mineral mining cartels that could also affect the geopolitical environment and future interstate warfare, like the Organization of the Petroleum Exporting Countries did in the 20th century for oil.[76] Eyl-Mazzega and Mathieu warn that

> [t]he concentration of resources in a small number of countries outside the OECD [Organisation for Economic Co-operation and Development], the oligopolistic nature of markets, and the fact that these resources are in the hands of powers which are often rivals (especially China and Russia) generate risks for access to resources . . . [and] could raise the total costs of the energy transition and block or threaten the development of national industries. This is especially so as competition is strengthen-

ing from military technologies, which are also big consumers of critical metals. Faced with trade tensions from the U.S., China may enhance its strategy of self-sufficiency and reinforce its pre-emption of resources.[77]

Månberger and Johansson explain the market dynamics behind this conceptual mechanism further, writing that

> resource rich countries may strive to secure incomes [from green minerals] that are often key to the
>
> national economies and state budgets. Depending on their positions, this can be carried out through attempts to control price setting and gain increased market shares. . . . These attempts can include measures that interfere– with the market and the interests of resource dependent countries, potentially leading to conflicts.[78]

We note, however, Månberger and Johansson's caveat that alarmist forms of this pathway should be tamped down because of the adaptability of market mechanisms:

> The geopolitical risks of the geographical concentration will depend on the availability of substitutes for use in renewable technologies, and the stability as well as geopolitical strategies of exporting countries. If countries choose to restrict supply or if antagonistic attacks and natural disasters in producer countries occur, it can result in physical shortages and price hikes. However, these are likely to have short-term impacts because substitution and alternative supply sources can alleviate shortages of a particular resource.[79]

On the other hand, we point out that a contradictory lesson could be drawn from the case of China's 2010 rare earth metals trade embargo discussed by Månberger and Johansson: specifically that 2010 demand levels for these metals were likely a fraction of what they will be in the future and that the short-lived move *nonetheless* engendered a global price spike.

Increased Interstate Competition for Scarce Resources

Hypothesis 6a: *Climate change–induced scarcity of renewable natural resources will provoke interstate simple-scarcity conflicts or zero-sum resource wars. Specifically, climate change–related scarcity of shared freshwater will increase tensions between nations straddling transboundary basins and lead to water wars.*

We reviewed approximately 30 prominent articles that investigated whether countries sharing water basins and river flows may be more likely to fight as a result of climate change.[80] Gleick presents the argument plainly: "Where water is scarce, competition for limited supplies can lead nations to see access to water as a matter of national security. History is replete with examples of competition and disputes over shared fresh water resources."[81] According to this argument, as freshwater scarcity or variability increases—whether because of higher mean temperatures, variable rainfall, or increased aridity—economic livelihoods and food security will be threatened; competing claims over transboundary rivers, lakes, and underground aquifers will increase; and countries may clash over shared water resources. The underlying theory is rooted in economic logic. The consumption of freshwater resources is adversarial and existential, supplies are capturable, demand is relatively inelastic in the short term, and substitutes are largely nonexistent in the short term. As a result, competition between water haves and have-nots may increase in the future, thereby raising hydro-political tensions that potentially escalate to low-level interstate conflict—or possibly full-out interstate war.

Our review of the interstate water war hypothesis literature found that the precise escalatory mechanism in this indirect pathway (i.e., the final determinative step) is somewhat fuzzy. Salehyan suggests that interstate war might result from bargaining failures when states use violence as a strategy to influence negotiations.[82] Spillmann explains that this realist logic essentially assumes that competition over water resources becomes "a zero-sum game."[83] He notes, "Because they can be seized, rivers are politically instrumentable. Upstream riparians can harness the water to put downstream states under pressure. For this reason, riparian disputes often commingle with traditional sources of conflict like territorial disputes or inter-state rivalries."[84] Hensel, McLaughlin Mitchell, and Sowers add a slightly finer point, specifying that scarcity-driven, zero-sum riparian interactions between states

> enhance the probability of conflict, especially if the downstream country is highly dependent on upstream water supply, the upstream country has the ability to seriously alter the quantity or quality of water flowing downstream, there is a history of antagonism between upstream and downstream states, and the downstream state is more powerful militarily.[85]

Institutional limitations on contract enforceability and authorities in the international system appear to be key to the mechanics of this pathway.[86] In Bächler's (1995) interpretation, this scarcity-driven pathway is driven by "the development dilemma and/or mismanagement, and not the classical security dilemma between nation states that is likely to lead to social tension, as well as political and even armed conflicts."[87] Scheffran, Link, and Schilling, on the other hand, argue that scarcity-driven "'security dilemma'" dynamics *could* "be triggered by climate change if threat perceptions are increased in times of crisis. . . . The degradation of natural resources puts the survival of people at stake, provoking the use of violence to protect key resources against competitors [rather than seeking to strengthen mutually beneficial cooperation]."[88] Homer-Dixon hypothesizes that climate change–induced crop failures could also lead agricultural and livestock exporters to weaponize scarce food exports.[89] Some of the literature vaguely proposes that climate change may "shift the balance of power between states either regionally or globally, producing instabilities that could lead to war."[90]

The literature also highlights that interstate disputes over shared water sources have been historically common in the CENTCOM AOR. Bernauer and Siegfried observe that the "Aral Sea basin [has] experienced international disputes over water allocation ever since the USSR collapsed," and they hypothesize that climate change–induced shifts in the runoff from the Syr Darya could drive a militarized interstate dispute between Uzbekistan and Kyrgyzstan in the medium to long term.[91] Gleick points out the frequency of past internationalized conflicts over access to the Nile, Jordan, and Euphrates rivers.[92] Podesta and Ogden comment that the "enormously intricate water politics of the [Middle East] have been aptly described as a 'hydropolitical security complex.' The Jordan River physically links the water interests of Israel, Jordan, Lebanon, the Palestinian Authority, and Syria; the Tigris and Euphrates Rivers physically link the interests of Iran, Iraq, Syria, and Turkey."[93] Scheffran and Battaglini similarly argue that "water has traditionally been a strategic issue in the Middle East, intertwined with the region's deeply rooted conflicts. . . . The arid climate, the imbalance between water demand and supply, and the ongoing confrontation between key political actors exacerbate the water crisis. Since much of the water resources are trans-boundary, water disputes often coincide with land disputes."[94]

Hypothesis 6b: *Climate change–related cross-border migration will exacerbate interstate rivalries via scarcity or security mechanisms, potentially escalating to grievance-based interstate war.*

As noted in the previous section on causal pathways from climate change to different forms of intrastate conflict, considerable research addresses how migration may serve as a link in the chain from climate change to conflict. Likewise, a large body of research examines pathways from *non-climate*–related migration to interstate crises and war, but relatively less work has been done to examine the potential link between climate change-related migration and interstate war. That said, some voices in the academic literature hypothesize about these pathways. These authors posit that climate change–related scarcity could lead to livelihood and food insecurity in developing countries that prompts waves of cross-border forced migration; this theory argues that migrant flows will increase the potential for competition over scarce jobs, resources, and assistance in the receiving countries and ultimately lead to host-nation grievances against migrant populations because of fears of upsetting the ethnic balance of power or fears of catalyzing separatist movements (i.e., among migrant communities that wish to reunify with their homeland).[95]

Cattaneo and Foreman contend that "international migration, especially from developing to developed countries and between developing countries, can significantly influence international relations, by eventually affecting the likelihood of interstate conflicts between receiving and sending countries. . . . Climatic stress in fact, can intensify emigration flows . . . and these flows may induce interstate disputes."[96] More specifically, these authors' logic is rooted in the rationalist theory of interstate conflict, by which they argue that mass flows of migrants may lower the opportunity costs of fighting and potentially decrease the likelihood of arriving at negotiated resolutions to interstate disputes:

> Leaders try to compromise and negotiate in the first instance. However, there are some factors that influence the utility of acting violently or affect the opportunity costs of entering in a conflict. For example,
>
> control of rival and excludable good has historically influenced the incidence of militarized conflicts . . .
>
> Migrants may compete with locals for jobs and scarce resources, and this can generate a sense of hostility among natives. Labour market concerns are a strong driver of opposition towards migrants in hosting societies . . . [generating] strong misperceptions of some migrants' characteristics . . . . These misperceptions can contribute to feeding the "emotional" threat, which can drive interstate conflicts by increasing bargaining failures. In this circumstance, the receiv-

ing country of the flows has low incentives to make concessions to avoid escalation toward war. These concerns can be exacerbated when the population flows occur too rapidly to be smoothly absorbed, as it might be the case for climate-induced migrants.[97]

Alternatively, Reuveny theorizes that migrant flows might be viewed by receiving states as a ploy by rival sending states to intentionally destabilize the receiving regime, thereby precipitating interstate crisis and conflict.[98] In Reuveny's formulation, environmental migration may "generate distrust between the area of the migration's origin and host area."[99] Salehyan posits a similar dynamic with a different outcome that traces increased refugee flows to increased militarized interstate disputes: specifically that, under some specific conditions, refugee-sending countries might launch military attacks on neighboring territories if migrants or rebels in the receiving country engage in cross-border violence, stoking tensions between states.[100]

Hypothesis 6c: *Climate change–related shifts in natural boundaries could lead to conflict that escalates into interstate war over territorial disputes.*

In research closely related to work on water wars, the literature hypothesizes that new interstate territorial disputes will arise—potentially escalating to war—because of changing natural boundaries, particularly when those boundaries are demarcated by rivers, glaciers, seas, or deserts. Distinct from the previous pathway, this hypothesis is not scarcity-driven. Lee et al. argue that extreme weather events may produce "disasters . . . [that] increase interstate conflict by disrupting established borders (e.g., floods altering river flows) and generating new territorial disputes."[101] Territorial disputes could lead states to conflict via bargaining failures. Alternatively, Lee and Tanaka specify that misperceptions, security dilemmas, and/or balance of power shifts could lead to interstate war from shifting natural boundaries.[102]

Similarly, new human access to and state control over previously unapproachable ecosystems and territories (e.g., the Arctic) may catalyze interstate conflict and, potentially, redraw geopolitical boundaries by maps rather than natural boundaries. As articulated in the 2022 *National Security Strategy* and in a vast body of academic literature, climate change has and will continue to make parts of the Arctic more accessible through major glacier melts, raising the specter of interstate competition and conflict in the region over new human access to otherwise scarce natural resources (e.g., oil, gas, fisheries).[103] Stated as a more generalized pathway: Climate change may open new territories and ecosystems to human activity, leading states to project power in new places to compete for first mover control over resource rents (renewables and nonrenewables) and territorial control.[104] In turn, according to this line of reasoning, interstate disputes over territorial boundaries and resource rights are bound to increase, potentially leading to escalatory interstate crises and war.[105] According to Markowitz's core proposition, the crucial mechanism in this pathway—which he posits could lead to "the Return of the Great Game" in the Arctic—is that

> some states have a stronger preference for projecting military power to seek territory as a source of wealth and rents . . . [specifically,] the more a state depends economically on extracting income from land, the
>
> more it will invest in projecting power to secure territory and resources. . . .
>
> The economic structure of the state, and more specifically the degree to which the economy is structured to generate income from resources, affects the government's preferences through two causal pathways. First, economic structure conditions the state's source of income and the rate of return from investing in securing resources versus producing goods and services. Second, economic structure influences the degree to which the state's governing coalition is composed of and captured by individuals or organizations from the resource sector. . . .
>
> Domestic political institutions influence the regime's preference for territory through a third causal pathway: they condition the regime's value for the political benefits associate with land rents. Compared to the profits from producing goods and services, land rents are easier for a regime to monopolize control over and deny to the political opposition. All regimes should value these political benefits, but autocrats should value them more.[106]

We note, of course, that this potential Arctic conflict pathway is likely to primarily affect CENTCOM's OAIs in indirect ways. However, glacial melting is creating tensions in the CENTCOM AOR that are leading to constant jockeying over, for example, Pakistan and India's border along the Siachen Glacier.

Increased Risks from Security Dilemmas, Diversionary Incentives, and Inadvertent Escalation

Hypothesis 7a: *The climate change–related weakening of state capacity could shift the underlying balance of power, leading to interstate war.*

Some scholars have hypothesized that states weakened by climate change–related disasters or scarcity (e.g., food shortages) might become more inviting targets for interstate aggression. Iyigun, Nunn, and Qian's study of the long-term historical record associated with climate change during the Little Ice Age (1400–1900 CE) posits that food and human insecurity mechanisms caused by climate and environmental change reduced populations and increased migration patterns in some regions, weakening state capacity and increasing state vulnerability to external aggression.[107] According to these authors, there are many potential, theoretical causal pathways that link an increase in the number of foreign invasions to historical climate change events, specifically through increased opportunities for victory—i.e., "due to a reduction in the cost of invasion because natural barriers, such as rivers or seas, froze over and allowed for easier troop movements" or because "belligerent neighbors sometimes viewed the weakening of state capacity caused by climate change as a good opportunity for invasion" or "the impoverished agricultural sector made it easy for governments to recruit soldiers"—or an increase in perceived returns from violence (i.e., "because reduced agricultural production increased demand for other sources of revenue, which incentivized governments to invade relatively fertile neighbors").[108]

Expectations that future climate change–related disasters might diminish state capacity could also render interstate war between rivals more likely, specifically by exacerbating inherent security commitment problems. As Bas and McLean explain from a game theory perspective, "in disaster-prone areas, actors' rational expectations about the likelihood and magnitude of potential future disasters can make conflict more likely," if the disaster effects are viewed as possibly shifting the balance of power between state rivals.[109] They write that

> when states anticipate differential effects from potential future shocks on their capabilities, the anticipation can generate commitment problems between states before shocks' arrival, and conflict may result. More broadly . . . in addition to actual disaster events, expected future disasters can be a source of conflict, and empirical analyses of the link between disasters and conflict should take disaster expectations into account.[110]

The operative mechanism in this theoretical pathway is that leaders in countries that are frequently prone to or are more susceptible to disaster-generated effects will likely anticipate that their state capabilities will be weakened in the future. This could be because of anticipated disruptions to export revenue-generating activities (from mines, oil and gas pipelines, railheads, and other transportation infrastructure) that translate to reduced state military financing or human capital losses in potential military manpower and the defense industrial base output. In turn, Bas and McLean argue that "if decision-makers fear that they will lose bargaining power as a result of a future disaster and will have to settle for a smaller share of resources, the decision-makers may prefer conflict in the present period [while it is still in a relatively strong position], as the existing literature on power shifts suggests."[111]

Hypothesis 7b: *Climate change–related extreme weather events could incentivize diversionary wars.*

Another line of argument in the literature posits that climate change–related extreme events and "rapid-onset disasters" could act as "political shocks that disrupt a rivalry relationship," with the disasters precipitating "diversionary conflict" between two states, especially in conflict-prone regions.[112] Several intermediate steps might link cause and effect. Lee et al. explain that these theoretical, intermediate steps in the pathway are essentially driven by domestic political pressures and civil unrest mechanisms, by which leaders may seize an opportunity to use "the disaster as propaganda to justify their war positions":

> Disasters often strain the state's capacity to provide security for its people; leaders who fail to prepare or respond can face domestic costs. To avoid potential removal from office, leaders have incentives to divert the public's attention away from poor disaster response by adopting a more aggressive foreign policy. Scapegoating may be easier for leaders in rivalries when an unexpected shock occurs, such as a rapid-onset disaster. We hypothesize that the time between militarized disputes is shortened when pairs of states experience rapid-onset disasters.[113]

Hypothesis 7c: *The climate change–related search for energy independence leads to the proliferation of civil nuclear programs and the increased risk of nuclear incidents that inadvertently escalate to interstate war.*

In combination with other intervening conditions (e.g., poor governance, high corruption, political instability, preexisting territorial disputes and/or interstate rivalries), an indirect pathway to interstate nuclear warfare might also be associated with the expansion of civil nuclear programs.[114] This pathway proceeding to nuclear interstate conflict could result from geopolitical miscalculations and/or inadvertent crisis escalation over "countries' ambitious nuclear energy programs—ostensibly designed to combat climate change—as 'cover' for developing nuclear programs with weaponization potential in mind."[115] Alternatively, the fact that the other pathways described in this report could lead to more conflict and global competition and/or destabilization (writ large) could, according to Bunn, propel states into a new nuclear arms race, again raising the associated risks of proliferation, accident, escalation, etc.[116]

More generally, Parthemore, Femia, and Werrell posit that climate change is likely to increase the physical danger to nuclear plants and sites and "transit routes critical to the security of nuclear facilities and materials" because of "temperature extremes, flooding, fires, hurricanes, and other disasters that various scenarios of climate change will likely bring . . . [including] new patterns of sea level rise, storm surge, and persistent flooding in areas where it was previously an occasional nuisance."[117] At the same time, climate

Figure 2.3. Causal Map of Hypothesized, Indirect Pathways to Interstate War

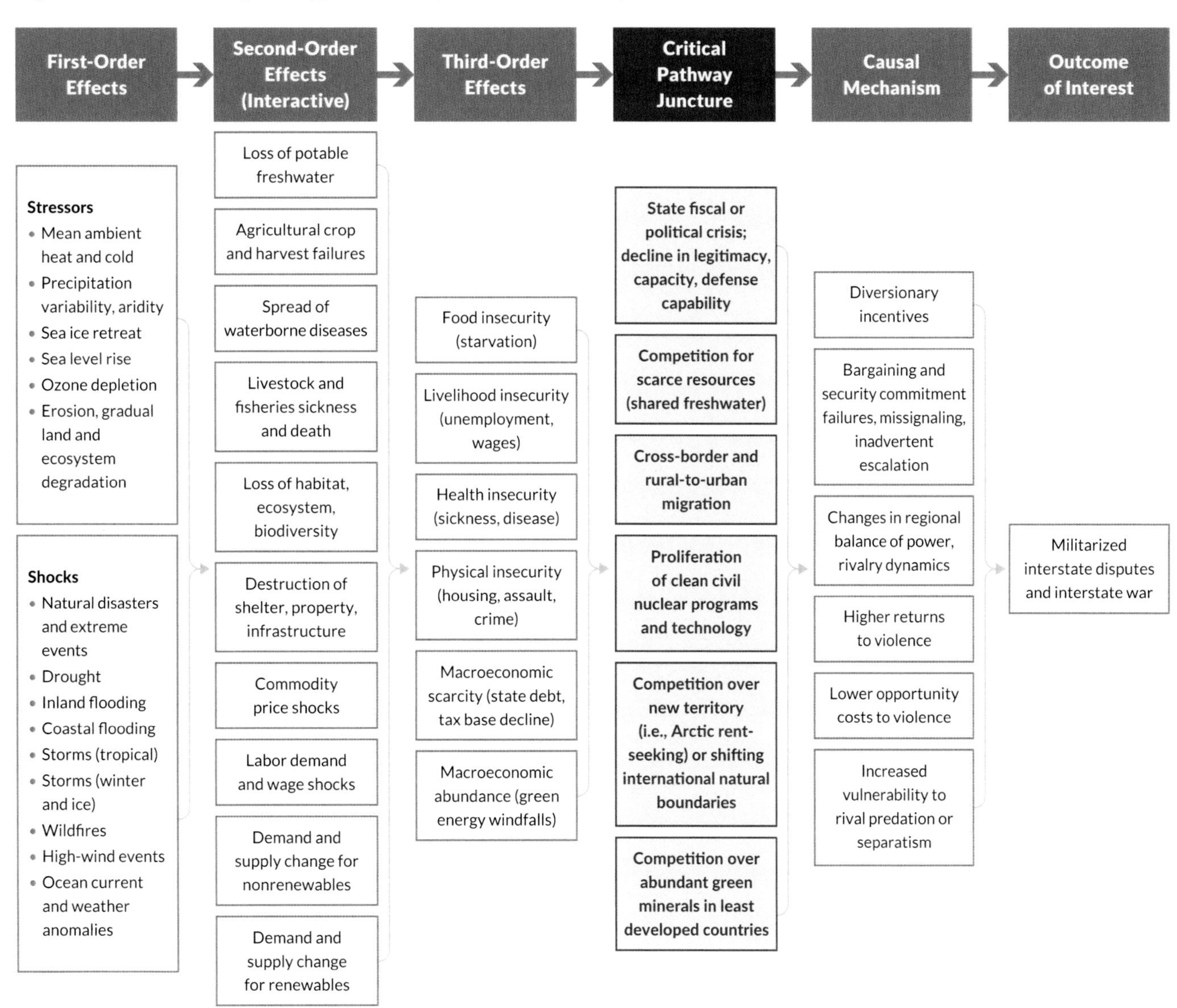

NOTE: This figure does not depict all the potential interactive effects across factors.

change will increase demand and competition for alternative energy sources (nongreenhouse gas emitting), including clean civil nuclear energy. This proliferation of civilian nuclear programs—plus the potential decrease in physical security from more powerful and more frequent storms—would logically increase the probability of nuclear accidents and/or the risk of insecure nuclear materiel, particularly between neighbors or regional rivals.

Figure 2.3 summarizes the causal pathways for interstate war.

Conclusion

Our review of the academic literature uncovered seven categories of causal pathways, and many of these pathways contain variants of more specific hypotheses that lead from climate change to conflict. The literature is most developed as it relates to the intrastate domain, which makes sense given the preponderance of this conflict in the international system. However, hypothesized causal pathways also exist from climate change to interstate conflict, even if this is a rarer phenomenon.

Intrastate pathways focus on climate change inducing resource curse dynamics via abundance mechanisms, intergroup or interpersonal competition for scarce resources, enhanced VNSA capacity and/or capability, and diminished state capacity and/or capability. The causal pathways we distilled for climate change contributing to interstate conflict also focus on resource curse dynamics and competition for scarce resources. However, the interstate domain includes climate change shifting the regional balance of power and incentivizing diversionary conflict, leading to inadvertent escalation.

Because nearly all of the causal pathways reviewed in this chapter are meant to apply globally, three real-world case studies are introduced in Chapter 3 to show how climate change has or could lead to conflict in the CENTCOM AOR. The case studies return to the causal pathways treated in Chapter 1 but highlight only the steps taken within the broader causal map that capture the real progression of the cases.

Endnotes

1 Bibliometrics refer to how often an article is cited in subsequent literature. The seven review articles are Thomas Bernauer, Tobias Böhmelt, and Vally Koubi, "Environmental Changes and Violent Conflict," *Environmental Research Letters*, Vol. 7, January 2012; Vally Koubi, Gabriele Spilker, Tobias Bohmelt, and Thomas Bernauer, "Do Natural Resources Matter for Interstate and Intrastate Armed Conflict?" *Journal of Peace Research*, Vol. 51, No. 2, March 2014; François Gemenne, Jon Barnett, W. Neil Adger, and Geoffrey Dabelko, "Climate and Security: Evidence, Emerging Risks, and a New Agenda," *Climate Change*, Vol. 123, No. 1, March 2014; Tobias Ide, "Research Methods for Exploring the Links Between Climate Change and Conflict," *WIREs Climate Change*, Vol. 8, No. 3, May–June January 2017; Vally Koubi, "Climate Change and Conflict," *Annual Review of Political Science*, Vol. 22, May 2019; Nina von Uexkull and Halvard Buhaug, "Security Implications of Climate Change: A Decade of Scientific Progress," *Journal of Peace Research*, Vol. 58, No. 1, January 2021; and Hans-Otto Pörtner, Debra C. Roberts, Melinda M. B. Tignor, Elvira Poloczanska, Katja Mintenbeck, Andrés Alegría, Marlies Craig, Stefanie Langsdorf, Sina Löschke, Vincent Möller, Andrew Okem, and Bardhyl Rama, eds., *Climate Change 2022: Impacts, Adaptation, and Vulnerability. Working Group II Contribution to the Sixth Assessment Report of the Intergovernmental Panel on Climate Change*, Intergovernmental Panel on Climate Change, Cambridge University Press, 2022.

2 Because the most recent of the seven review articles is from 2021, we supplemented our list with approximately 50 articles from 2021 to 2022, identified from the top journals on climate change and conflict. As warranted, we also reviewed numerous older, highly cited, peer-reviewed works not on the initial list.

3 The sample reviewed cannot be easily described by summary stats broken down by these conflict type categories. Many articles discuss pathways to more than one type of conflict, and many others did not clearly define the level of conflict under investigation or interchangeably used such terms as *armed conflict* or *civil conflict*.

4 For every article, readers coded some 20 data points of interest (e.g., independent and dependent variables considered, sample descriptions, key findings, nonproximate effects) and distilled any potential climate-conflict causal pathways and important findings on the relationship between climate hazards and conflict.

5 Koubi, 2019, p. 346.

6 While direct pathways are less germane to CENTCOM leaders and defense planners than indirect pathways are, they nonetheless hold relevance for future OAIs in the AOR. The appendix a contains a brief summary of the four potential one-step pathways to violence or conflict identified in our literature review.

7 Kristy H. Lewis and Timothy M. Lenton, "Knowledge Problems in Climate Change and Security Research," *WIREs Climate Change*, Vol. 6, No. 4, July–August 2015, p. 5.

8 For a good summary of the theoretical principles in the resource curse literature, see Henrik Urdal, "People vs. Malthus: Population Pressure, Environmental Degradation, and Armed Conflict Revisited," *Journal of Peace Research*, Vol. 42, No. 4, July 2005; Joshua D. Angrist and

Adriana D. Kugler, "Rural Windfall or a New Resource Curse? Coca, Income, and Civil Conflict in Colombia," *Review of Economics and Statistics*, Vol. 90, No. 2, May 2008.

9 *Green minerals* refer to minerals used in the production of renewable energy. The list of those materials grows with the development of new technologies but often includes eodymium, dysprosium, copper, lithium, cobalt, manganese, nickel, platinum, tellurium, indium, gallium, selenium, silver and silicon. See André Månberger and Bengt Johansson, "The Geopolitics of Metals and Metalloids Used for the Renewable Energy Transition," Energy Strategy Reviews, Vol. 26, November 2019

10 See Månberger and Johansson, 2019; Clare Church and Alec Crawford, "Minerals and the Metals for the Energy Transition: Exploring the Conflict Implications for Mineral-Rich, Fragile States," in Manfred Hafner and Simone Tagliapietra, eds., *The Geopolitics of the Global Energy Transition*, Lecture Notes in Energy, Vol. 73, Springer, 2020; and Roger E. Meiners and Andrew P. Morriss, *Addressing Green Energy's "Resource Curse,"* Texas A&M University School of Law, No. 22–31, February 16, 2022.

11 Church and Crawford, 2020, p. i.

12 Jon Barnett, "Security and Climate Change," *Global Environmental Change*, Vol. 13, No. 1, April 2003, p. 11.

13 Lasse Eisgruber, "The Resource Curse: Analysis of the Applicability to the Large-Scale Export of Electricity from Renewable Resources," *Energy Policy*, Vol. 57, June 2013; André Månsson, "A Resource Curse for Renewables? Conflict and Cooperation in the Renewable Energy Sector," *Energy Research and Social Science*, Vol. 10, November 2015; Scott W. Lyons, "Preventing a Renewable Resource Curse," *Sustainable Development Law and Policy*, Vol. 15, No. 2, Spring 2015; Alycia Leonard, Aniq Ahsan, Flora Charbonnier, and Stephanie Hirmer, "The Resource Curse in Renewable Energy: A Framework for Risk Assessment," *Energy Strategy Reviews*, Vol. 41, May 2022.

14 Lyons, 2015, p. 4.

15 As Baechler argues, "The catalysts to escalate center-periphery conflicts are primarily large cash crop farming projects, dams, and mining" (Günther Baechler, "Why Environmental Transformation Causes Violence: A Synthesis," *Environmental Change and Security Project Report*, Wilson Center, No. 4, Spring 1998, p. 27).

16 A distinct field of research debates the ways in which Malthus's preindustrial, dystopian predictions have (or have not) been proven spectacularly wrong. While a review of this related body of literature is beyond the scope of this report, we note that the most commonly cited criticisms focus on Malthus's failure to foresee the transformative role of technology in the advancement of human food production, health care and life expectancy, extractive capabilities, global trade patterns, etc. In short, Malthus overpredicted population growth rates and demographic pressures and underpredicted human coping mechanisms (Urdal, 2005, p. 418).

17 Thomas Malthus, *An Essay on the Principle of Population*, London, 1798.

18 Urdal, 2005, p. 418.

19 As mediating conditions, this logic continues to assert that countries with higher population growth rates (especially urban, refugee, and migrant populations) and/or population density may be more likely to experience intrastate war (Urdal, 2005, p. 418).

20 Urdal, 2005, pp. 418–420.

21 While the neo-Malthusian school advanced by Homer-Dixon et al. still retains currency, it has also been robustly challenged. Salehyan argues that these hypotheses are overly deterministic. He writes,

> The overly structural logic linking climate change to armed conflict ignores human agency, ingenuity, the potential for technological innovation, and the vital role of political institutions in managing conflict (or failing to do so). . . . Nonetheless, scholars making strong claims about the primacy of environmental conditions continue to argue that human ingenuity and the redistributive functions of governments are likely to be overwhelmed by environmental stress (Idean Salehyan, "From Climate Change to Conflict? No Consensus Yet," *Journal of Peace Research*, Vol. 45, No. 3, May 2008, pp. 315, 318).

22 Thomas Homer-Dixon, "On the Threshold: Environmental Changes as Causes of Acute Conflict," *International Security*, Vol. 16, No. 2, Fall 1991; Thomas F. Homer-Dixon, "Environmental Scarcities and Violent Conflict: Evidence from Cases," *International Security*, Vol. 19, No. 1, Summer 1994; Thomas Homer-Dixon, "The Project on Environment, Population and Security: Key Findings of Research," *Environmental Change and Security Project Report 2*, Woodrow Wilson Center, 1996; Thomas Homer-Dixon and Jessica Blitt, "Introduction: A Theoretical Overview," in Thomas Homer-Dixon and Jessica Blitt, eds., *Ecoviolence: Links Among Environment, Population, and Security*, Rowman and Littlefield, 1998; Thomas F. Homer-Dixon, *Environment, Scarcity, and Violence*, Princeton University Press, 1999. See also Nils Petter Gleditsch and Henrik Urdal, "Ecoviolence? Links Between Population Growth, Environmental Scarcity, and Violent Conflict in Thomas Homer-Dixon's Work," *Journal of International Affairs*, Vol. 56, No. 1, Fall 2002.

23 Homer-Dixon, 1996, p. 46.

24 Koubi, 2019, p. 346.

25 Patrick Meier, Doug Bond, and Joe Bond, "Environmental Influences on Pastoral Conflict in the Horn of Africa," *Political Geography*, Vol. 26, No. 6, August 2007; Ayalneh Bogale and Benedikt Korf, "To Share or Not to Share? (Non-)Violence, Scarcity, and Resource Access in Somali Region, Ethiopia," *Journal of Development Studies*, Vol. 43, No. 4, May 2007; Clionadh Raleigh and Dominic Kniveton, "Come Rain or Shine: An Analysis of Conflict and Climate Variability in East Africa," *Journal of Peace Research*, Vol. 49, No. 1, January 2012; Karen M. Witsenburg and Wario R. Andano, "Of Rain and Raids: Violent Livestock Raiding in Northern Kenya," *Civil Wars*, Vol. 11, No. 4, December 2009; Carol R. Ember, Teferi Abate Adem, Ian Skoggard, and Eric C. Jones, "Livestock Raiding and Rainfall Variability in Northwestern Kenya," *Civil Wars*, Vol. 14, No. 2, June 2012; Hanne Fjelde and Nina von Uexkull, "Climate Triggers: Rainfall Anomalies, Vulnerability and Communal Conflict in Sub-Saharan Africa," *Political Geography*, Vol. 31, No. 7, September 2012; Christopher K. Butler and Scott Gates, "African Range Wars: Climate, Conflict, and Property Rights," *Journal of Peace Research*, Vol. 49, No. 1, January 2012; Ole Magnus Theisen, "Climate Clashes? Weather Variability, Land Pressure, and Organized Violence in Kenya, 1989–2004," *Journal of Peace Research*, Vol. 49, No. 1, January 2012; Tor A. Benjaminsen, Koffi Alinon, Halvard Buhaug, and Jill Tove Buseth, "Does Climate Change Drive Land-Use Conflicts in the Sahel?" *Journal of Peace Research*, Vol. 49, No. 1, January 2012; Carol R. Ember, Ian Skoggard, Teferi Abate Adem, and A. J. Faas, "Rain and Raids Revisited: Disaggregating Ethnic Group Livestock Raiding in the Ethiopian-Kenyan Border Region," *Civil Wars*, Vol. 16, No. 3, July 2014; Kostadis J. Papaioannou, "Climate Shocks and Conflict: Evidence from Colonial Nigeria," *Political Geography*, Vol. 50, January 2016.

26 Günther Bächler, "The Anthropogenic Transformation of the Environment: A Source of War?" in Kurt R. Spillmann and Günther Bächler (eds.), *Environmental Crisis: Regional Conflicts and Ways of Cooperation, Environment and Conflicts Project*, No. 14, September 1995, p. 15.

27 Meier, Bond, and Bond, 2007, p. 722.

28 Jonas Vestby, "Climate Variability and Individual Motivations for Participating in Political Violence," *Global Environmental Change*,

Vol. 56, May 2019, p. 114. See also Jon Barnett and W. Neil Adger, "Climate Change, Human Security, and Violent Conflict," *Political Geography*, Vol. 26, No. 6, August 2007; and Koubi, 2019, p. 347.

29 Todd Graham Smith, "Feeding Unrest: Disentangling the Causal Relationship Between Food Price Shocks and Sociopolitical Conflict in Urban Africa," *Journal of Peace Research*, Vol. 51, No. 6, November 2014, p. 682.

30 Ravnborg et al. (2012) finds that violence can occur at the local level as a result of competition over water resources but that this is rare and that cooperative cases are equally common (Helle Munk Ravnborg, Rocio Bustamante, Abdoulaye Cissé, Signe M. Cold-Ravnkilde, Vladimir Cossio, Moussa Djiré, Mikkel Funder, Ligia I. Gómez, Phuong Le, Carol Mweemba, Imasiku Nyambe, Tania Paz, Huong Pham, Roberto Rivas, Thomas Skielboe, and Nguyen T. B. Yen, "Challenges of Local Water Governance: The Extent, Nature and Intensity of Local Water-Related Conflict and Cooperation," *Water Policy*, Vol. 14, No. 2, April 2012).

31 Jürgen Scheffran and Antonella Battaglini, "Climate and Conflicts: The Security Risks of Global Warming," *Regional Environmental Change*, Vol. 11, Supp. 1, March 2010, p. 31.

32 Adam Yeeles, "Weathering Unrest: The Ecology of Urban Social Disturbances in Africa and Asia," *Journal of Peace Research*, Vol. 52, No. 2, March 2015, p. 159.

33 Nils Petter Gleditsch, "This Time Is Different! Or Is It? NeoMalthusians and Environmental Optimists in the Age of Climate Change," *Journal of Peace Research*, Vol. 58, No. 1, January 2021, pp. 181–182.

34 Fjelde and von Uexkull, 2012, p. 444. For an alternative view challenging the strength of this argument, see Benjaminsen et al., 2012.

35 Jan Selby and Clemens Hoffmann, "Beyond Scarcity: Rethinking Water, Climate Change, and Conflict in the Sudans," *Global Environmental Change*, Vol. 29, November 2014.

36 Tobias Ide, Michael Brzoska, Jonathan F. Donges, and Carl-Friedrich Schleussner, "Multi-Method Evidence for When and How Climate-Related Disasters Contribute to Armed Conflict Risk," *Global Environmental Change*, Vol. 62, May 2020, p. 2.

37 Ide et al., 2020, p. 2.

38 See, for example, Michael Brzoska and Christiane Fröhlich, "Climate Change, Migration, and Violent Conflict: Vulnerabilities, Pathways, and Adaptation Strategies," *Migration and Development*, Vol. 5, No. 2, 2016; Vally Koubi, Tobias Böhmelt, Gabriele Spilker, and Lena Schaffer, "The Determinants of Environmental Migrants' Conflict Perception," *International Organization*, Vol. 72, No. 4, Fall 2018; Rafael Reuveny, "Ecomigration and Violent Conflict: Case Studies and Public Policy Implications," *Human Ecology*, Vol. 36, No. 1, February 2008; Justin Schon and Stephen Nemeth, "Moving into Terrorism: How Climate-Induced Rural-Urban Migration May Increase the Risk of Terrorism," *Terrorism and Political Violence*, Vol. 34, No. 5, June 2022; Andrea Malji, Laurabell Obana, and Cidney Hopkins, "When Home Disappears: South Asia and the Growing Risk of Climate Conflict," *Terrorism and Political Violence*, Vol. 34, No. 5, May 2022.

39 Koubi, 2019, p. 348; Rafael Reuveny, "Climate Change-Induced Migration and Violent Conflict," *Political Geography*, Vol. 26, 2007.

40 Ide et al., 2020, p. 2.

41 Matt McDonald, "Discourses of Climate Security," *Political Geography*, Vol. 33, March 2013, p. 47.

42 Bogale and Korf, 2007, p. 744.

43 Christiane Fröhlich and Giovanna Gioli, "Gender, Conflict, and Global Environmental Change," *Peace Review*, Vol. 27, No. 2, 2015, p. 137. See also Belén Sanz-Barbero, Cristina Linares, Carmen Vives-Cases, José Luis González, Juan José López-Ossorio, Julio Díaz, "Heat Wave and the Risk of Intimate Partner Violence," *Science of the Total Environment*, Vol. 644, December 2018.

44 Huong Thu Nguyen, "Gendered Vulnerabilities in Times of Natural Disasters: Male-to-Female Violence in the Philippines in the Aftermath of Super Typhoon Haiyan," *Violence Against Women*, Vol. 25, No. 4, August 2018, p. 421.

45 Military Advisory Board, *National Security and the Threat of Climate Change*, CNA Corporation, 2007, p. 16.

46 Zorzeta Bakaki and Roos Haer, "The Impact of Climate Variability on Children: The Recruitment of Boys and Girls by Rebel Groups," *Journal of Peace Research*, Vol. 60, No. 4, July 2023; Oeindrila Dube and Juan F. Vargas, "Commodity Price Shocks and Civil Conflict: Evidence from Colombia," *Review of Economic Studies*, Vol. 80, No. 4, October 2013, pp. 1384–1385; Vestby, 2019, p.114.

47 Vestby, 2019, p. 114.

48 Alison Heslin, "Riots and Resources: How Food Access Affects Collective Violence," *Journal of Peace Research*, Vol. 58, No. 2, April 2020, p. 209–210.

49 Ashton Kingdon and Briony Gray, "The Class Conflict Rises When You Turn up the Heat: An Interdisciplinary Examination of the Relationship between Climate Change and Left-Wing Terrorist Recruitment," *Terrorism and Political Violence*, Vol. 34, No. 5, May 2022.

50 Kate Burrows and Patrick L. Kinney, "Exploring the Climate Change, Migration, and Conflict Nexus," *International Journal of Environmental Research and Public Health*, Vol. 13, No. 4., April 2016, p. 8.

51 Burrows and Kinney, 2016, p. 8.

52 Reuveny, 2008, p. 659.

53 Paul J. Smith, "Climate Change, Mass Migration, and the Military Response," *Orbis*, Vol. 51, No. 4, 2007, p. 620.

54 Reuveny, 2008, p. 659.

55 Schon and Nemeth, 2022, p. 933.

56 Colin H. Kahl, *States, Scarcity, and Civil Strife in the Developing World*, Princeton University Press, 2006.

57 Homer-Dixon, 1994, pp. 6–7. Numerous rival voices in the literature criticize these theories that focus on the role of the state as too simplistic, particularly as they ignore cooperative and adaptive coping mechanisms, which are beyond the scope of this analysis. See Gemenne et al., 2014, p. 2. See also Salehyan, 2008, p. 320.

58 Thomas Homer-Dixon, "Terror in the Weather Forecast," *New York Times*, April 24, 2007.

59 Halvard Buhaug, Tor A. Benjaminsen, Espen Sjaastad, and Ole Magnus Theisen, "Climate Variability, Food Production Shocks, and Violent Conflict in Sub-Saharan Africa," *Environmental Research Letters*, Vol. 10, No. 12, December 2015, pp. 2–3.

60 Alan Dupont and Graeme Pearman, *Heating Up the Planet: Climate Change and Security*, Lowy Institute, Paper No. 12, 2006, p. viii.

61 Barnett and Adger, 2007, p. 651.

62 Cullen S. Hendrix and Idean Salehyan, "Climate Change, Rainfall, and Social Conflict in Africa," *Journal of Peace Research*, Vol. 49, No. 1, January 2012, p. 38; Military Advisory Board, 2007, pp. 15–18; Heslin, 2020, pp. 209–211.

63 Dube and Vargas, 2013; Vestby, 2019, p. 114; Benjamin T. Jones, Eleonora Mattiacci, and Bear F. Braumoeller, "Food Scarcity and State Vulnerability: Unpacking the Link Between Climate Variability and Violent Unrest," *Journal of Peace Research*, Vol. 54, No. 3, May 2017, p. 339.

64 Mark Pelling and Kathleen Dill, "Disaster Politics: Tipping Points for Change in the Adaptation of Sociopolitical Regimes," *Progress in Human Geography*, Vol. 34, No. 1, February 2010, p. 22.

65 Pelling and Dill, 2010, p. 22.

66 Pelling and Dill, 2010, p. 22.

67 Joshua Eastin, "Fuel to the Fire: Natural Disasters and the Duration of Civil Conflict," *International Interactions*, Vol. 42, No. 2, February 2016.

68 See, for example, Peter H. Gleick, "Water, Drought, Climate Change, and Conflict in Syria," *Weather, Climate, and Society*, Vol. 6, No. 3, July 2014; Francesca De Châtel, "The Role of Drought and Climate Change in the Syrian Uprising: Untangling the Triggers of the Revolution," *Middle Eastern Studies*, Vol. 50, No. 4, May 2014; Colin P. Kelley, Shahrzad Mohtadi, Mark A. Cane, Richard Seager, and Yochanan Kushnir, "Climate Change in the Fertile Crescent and Implications of the Recent Syrian Drought," *Proceedings of the National Academy of Sciences*, Vol. 112, No. 11, March 17, 2015; Konstantin Ash and Nick Obradovich, "Climatic Stress, Internal Migration, and Syrian Civil War Onset," *Journal of Conflict Resolution*, Vol. 64, No. 1, January 2020, pp. 3–31; Jan Selby, Omar S. Dahi, Christiane Fröhlich, and Mike Hulme, "Climate Change and the Syrian Civil War Revisited," *Political Geography*, Vol. 60, September 2017; Arnon Karnieli, Alexandra Shtein, Natalya Panov, Noam Weisbrod, and Alon Tal, "Was Drought Really the Trigger Behind the Syrian Civil War in 2011?" *Water*, Vol. 11, No. 8, July 2019; and Andrew M. Linke and Brett Ruether, "Weather, Wheat, and War: Security Implications of Climate Variability for Conflict in Syria," *Journal of Peace Research*, Vol. 58, No. 1, January 2021.

69 For a similar finding, see Kendra Sakaguchi, Anil Varughese, and Graeme Auld, "Climate Wars? A Systematic Review of Empirical Analyses on the Links Between Climate Change and Violent Conflict," *International Studies Review*, Vol. 19, No. 4, December 2017, p. 628.

70 Bernauer, Böhmelt, and Koubi, 2012; Urdal, 2005, p. 420. See also Dupont and Pearman, 2006, p. viii; and Homer-Dixon, 1999, p. 5.

71 As Bernauer, Böhmelt, and Koubi point out, the empirical evidence on these links generally supports hypotheses that argue climate change will lead to lower-level interstate conflict rather than full-blown interstate war (Bernauer, Böhmelt, and Koubi, 2012).

72 Hendrix posits that high oil prices may embolden oil-rich states to behave more aggressively, leading to increased interstate disputes among petrostates when prices rise. See Cullen S. Hendrix, "Oil Prices and Interstate Conflict," *Conflict Management and Peace Science*, Vol. 34, No. 6, November 2017.

73 Benjamin K. Sovacool and Walter Götz, "Major Hydropower States, Sustainable Development, and Energy Security: Insights from a Preliminary Cross-Comparative Assessment," *Energy*, Vol. 142, January 2018, p. 1075.

74 Paul R. Hensel, Sara McLaughlin Mitchell, and Thomas E. Sowers II, "Conflict Management of Riparian Disputes," *Political Geography*, Vol. 25, No. 4, May 2006, p. 391.

75 For one of the first takes on this now well-understood pathway connecting geopolitical risk to green technology supply chain vulnerabilities and monopolies and oligopolies, see Komal Habib, Lorie Hamelin, and Henrik Wenzel, "A Dynamic Perspective of the Geopolitical Supply Risk of Metals," *Journal of Cleaner Production*, Vol. 133, October 2016, pp. 850–858. See also Church and Crawford, 2020, pp. 279–304.

76 Megan O'Sullivan, Indra Overland, and David Sandalow, *The Geopolitics of Renewable Energy*, Columbia University Center on Global Energy Policy and Harvard Kennedy School Belfer Center for Science and International Affairs, June 2017, pp. 11–12.

77 Marc-Antoine Eyl-Mazzega and Carole Mathieu, "The European Union and the Energy Transition," in Manfred Hafner and Simone Tagliapietra, eds., *The Geopolitics of the Global Energy Transition*, Lecture Notes in Energy, Vol. 73, Springer, 2020, p. 40.

78 Månberger and Johansson, 2019, p. 2.

79 Månberger and Johansson, 2019, p. 8.

80 In the words of Spillmann, "Water shortage is the environmental problem number one and most prone to lead to violent conflicts or war." See Kurt R. Spillmann, "From Environmental Change to Environmental Conflict," in Kurt R. Spillmann and Günther Bächler, *Environmental Crisis: Regional Conflicts and Ways of Cooperation*, Center for Security Studies, Environment and Conflicts Project, No. 14, September 1995, p. 8. For other prominent, early work in the field, see Peter H. Gleick, "Water and Conflict: Fresh Water Resources and International Security," *International Security*, Vol. 18, No. 1, Summer 1993; Homer-Dixon, 1994, pp. 6–7; Hans Petter Wollebæk Toset, Nils Petter Gleditsch, and Håvard Hegre, "Shared Rivers and Interstate Conflict," *Political Geography*, Vol. 19, No. 8, November 2000; Shira Yoffe, Aaron T. Wolf, and Mark Giordano, "Conflict and Cooperation over International Freshwater Resources: Indicators of Basins at Risk," *Journal of the American Water Resources Association*, Vol. 39, No. 5, October 2003; Nils Petter Gleditsch, Kathryn Furlong, Håvard Hegre, Bethany Lacina, and Taylor Owen, "Conflicts over Shared Rivers: Resource Scarcity or Fuzzy Boundaries?" *Political Geography*, Vol. 25, No. 4, May 2006; Hensel, McLaughlin Mitchell, and Sowers, 2006.

Many scholars question the simple narrative of the so-called water wars; however, as Bernauer and Böhmelt argue, "Studies . . . explaining armed conflict or militarized disputes in terms of water stress, have produced inconclusive findings. Even if there is evidence for some water-related influence, other determinants of armed conflict actually play a much more important role than water stress" (Thomas Bernauer and Tobias Böhmelt, "International Conflict and Cooperation over Freshwater Resources," *Nature Sustainability*, Vol. 3, No. 5, 2020, p. 351); A conflicting liberal institutionalist strain in the literature asks whether acute water scarcity might encourage peace and cooperation rather than neo-Malthusian competition and conflict, particularly in the short run. Brochmann and Gleditsch contend that both views are "spurious" and that it is impossible to "establish a conflict-inducing effect of shared rivers over and beyond contiguity itself . . . [given that] nearly all neighbors in the international system share at least one river" (Marit Brochmann and Nils Petter Gleditsch, "Shared Rivers and Conflict—A Reconsideration," *Political Geography*, Vol. 31, No. 8, November 2012, p. 519); see Gemenne et al., 2014, pp. 5–6; Anna Kalbhenn, "Liberal Peace and Shared Resources—A Fair-Weather Phenomenon?" *Journal of Peace Research*, Vol. 48, No. 6, November 2011; Shlomi Dinar, David Katz, Lucia De Stefano, and Brian Blankespoor, "Climate Change, Conflict, and Cooperation: Global Analysis of the Effectiveness of International River Treaties in Addressing Water Variability," *Political Geography*, Vol. 45, March 2015; Giorgos Kallis and Christos Zografos, "Hydro-Climatic Change, Conflict and Security," *Climatic Change*, Vol. 123, No. 1, March 2014; and Colleen Devlin and Cullen S. Hendrix, "Trends and Triggers Redux: Climate Change, Rainfall, and Interstate Conflict," *Political Geography*, Vol. 43, November 2014.

81 Gleick, 1993, p. 79.

82 Salehyan, 2008, p. 317.

83 Spillmann, 1995, p. 8.

84 Spillmann, 1995, p. 8.

85 Hensel, McLaughlin Mitchell, and Sowers, 2006, p. 388.

86 Hensel, McLaughlin Mitchell, and Sowers, 2006, pp. 389–390.

87 Bächler, 1995, p. 11.

88 Jürgen Scheffran, P. Michael Link, and Janpeter Schilling, "Theories and Models of the Climate-Security Interaction: Framework and Application to a Climate Hot Spot in North Africa," in Jürgen Scheffran, Michael Brzoska, Hans Günter Brauch, Peter Michael Link,

and Janpeter Schilling, eds., *Climate Change, Human Security, and Violent Conflict: Challenges for Societal Stability*, Springer, 2012, pp. 101–102.

[89] Homer-Dixon, 1991, pp. 44–45. See also Peter Wallensteen, "Food Crops as a Factor in Strategic Policy and Action," in Arthur H. Westing, ed., *Global Resources and International Conflict: Environmental Factors in Strategic Policy and Action*, Oxford University Press, 1986.

[90] Homer-Dixon, 1991, p. 44.

[91] Thomas Bernauer and Tobias Siegfried, "Climate Change and International Water Conflict in Central Asia," *Journal of Peace Research*, Vol. 49, No. 1, January 2012, p. 227.

[92] Gleick, 1993, p. 79.

[93] John Podesta and Peter Ogden, "The Security Implications of Climate Change," *Washington Quarterly*, Vol. 31, No. 1, January 2008, p. 121.

[94] According to these authors, "Most contentious has been the sharing of the water of the Jordan River basin among Israel, Jordan, Lebanon, Syria, and the Palestinians, raising issues of equity. . . . Israel receives more than half of its water resources from occupied Arab territories and has a higher per-capita consumption than its neighbours. Reduced water supply over an extended period also bears a conflict potential among the countries in the Nile basin. Particularly, Egypt depends on the Nile for 95 percent of its drinking and industrial water and could feel threatened by countries upstream that exploit water from the river" (Scheffran and Battaglini, 2011, pp. 34–35).

[95] Reuveny, 2007, p. 659. See also Reuveny, 2008.

[96] Cristina Cattaneo and Timothy Foreman, *Climate Change, International Migration, and Interstate Conflict*, Centre for Research and Analysis of Migration, CDP 09/21, March 31, 2021, p. 2.

[97] Cattaneo and Foreman, 2021, pp. 2–3.

[98] Reuveny, 2008, p. 659.

[99] Reuveny, 2008, p. 659.

[100] Salehyan cites as theoretical examples (although they are not related to climate change) the cases of Israel's invasion of Lebanon in 1982 and Rwanda's invasion of the Democratic Republic of the Congo in 1996. See Salehyan, 2008, p. 788.

[101] Bomi K. Lee, Sara McLaughlin Mitchell, Cody J. Schmidt, and Yufan Yang, "Disasters and the Dynamics of Interstate Rivalry," *Journal of Peace Research*, Vol. 59, No. 1, March 2022.

[102] James R. Lee and Kisei R. Tanaka, "Climate Change, Conflict, and Moving Borders," *International Journal of Climate Change Impacts and Responses*, Vol. 8, No. 3, January 2016.

[103] Joseph R. Biden, *National Security Strategy*, White House, October 2022. For a few of the earliest warnings about this potential pathway, see Scott G. Borgerson, "Arctic Meltdown: The Economic and Security Implications of Global Warming," *Foreign Affairs*, Vol. 87, No. 2, March–April 2008; Oran R. Young, "Whither the Arctic? Conflict or Cooperation in the Circumpolar North," *Polar Record*, Vol. 45, No. 1, January 2009; and Kristian Åtland, *Security Implications of Climate Change in the Arctic*, Norwegian Defence Research Establishment, No. 2010/01097, May 2010.

[104] Rolf Tamnes and Kristine Offerdal, "Introduction," in Rolf Tamnes and Kristine Offerdal, eds., *Geopolitics and Security in the Arctic: Regional Dynamics in a Global World*, Routledge, 2014, pp. 1–11.

[105] We note that other voices in the literature, such as Young, regard it as healthy to consider "the possibility of extreme events, regardless of the likelihood of their occurrence . . . [including] scary scenarios regarding the future of the Arctic," but they nonetheless characterize scenarios predicting interstate war over Arctic resources to be "far fetched" (Young, 2009, p. 74).

[106] Jonathan N. Markowitz, *Perils of Plenty: Arctic Resource Competition and the Return of the Great Game*, Oxford University Press, 2020, pp. 4–6.

[107] See Murat Iyigun, Nathan Nunn, and Nancy Qian, *Winter is Coming: The Long-Run Effects of Climate Change on Conflict, 1400–1900*, IZA Institute of Labor Economics, No. 10475, January 2017.

[108] Iyigun, Nunn, and Qian, 2017, pp. 11–12.

[109] Muhammet A. Bas and Elena V. McLean, "Expecting the Unexpected: Disaster Risks and Conflict," *Political Research Quarterly*, Vol. 74, No. 2, June 2021, p. 421.

[110] Again, these authors recognize that disasters may, on the other hand, have a pacifying effect on interstate rivalries (Bas and McLean, 2021, p. 422).

[111] Bas and McLean, 2021, p. 423.

[112] These authors acknowledge that climate-related disasters could alternatively lead to peace, cooperation, and rapprochement between interstate rivals. See Lee et al., 2022, p. 12.

[113] Lee et al., 2022, p. 12.

[114] Gemenne et al., 2014, p. 5.

[115] We note that this pathway may have an important real-life application in the CENTCOM AOR via Pakistan's fears of India's civil nuclear ambitions. As Parthemore, Femia, and Werrell explain,

> Nuclear and climate connections are reaching into the very fabric of international relations as well. For several years, India has cited the need for expanded use of nuclear power worldwide to reduce the emissions of carbon dioxide that drive climate change as one reason it should be admitted to the Nuclear Suppliers Group and be able to trade freely in nuclear technology—bringing another difficult dynamic into an already complicated jumble of political and security issues in that forum (Christine Parthemore, Francesco Femia, and Caitlin Werrell, "The Global Responsibility to Prepare for Intersecting Climate and Nuclear Risks," *Bulletin of the Atomic Scientists*, Vol. 74, No. 6, 2018, p. 376).

[116] Matthew Bunn, "Nuclear Disarmament, Nuclear Energy, and Climate Change: Exploring the Linkages," in Bård Nikolas Vik Steen and Olav Njølstad, eds., *Nuclear Disarmament: A Critical Assessment*, Routledge, 2019.

[117] While the 2011 Fukushima Daiichi nuclear power plant accident was caused by an earthquake-related tsunami—not a climate-related extreme event—it provided a wake-up call about the need to harden the world's nuclear power plant facilities, waste storage, and research and development sites (Parthemore, Femia, and Werrell, 2018, p. 375).

CHAPTER 3

ILLUSTRATIVE CASE STUDIES OF CAUSAL PATHWAYS

AS DISCUSSED IN THE literature review, climate change can contribute to different types of insecurity: conflict below the threshold of war, intrastate war, and interstate war. Although the literature we reviewed was global in scope, the causal pathways identified in the literature have already occurred in the CENTCOM AOR or have the potential to occur in the future. This chapter presents three short case studies as illustrations of how these causal pathways are already unfolding in the CENTCOM AOR. In keeping with the overarching observation in the literature in Chapter 2 that climate change is "never a sole or sufficient cause of large migrations, poverty, or violence; it always joins with other economic political and social factors to produce its effects," none of the profiled conflicts are solely caused by climate hazards.[1] Rather, the analysis shows how climate hazards interact with other factors to contribute to the onset or intensification of existing conflicts.

Protests in Basra as an Illustration of Intrastate Conflict Below the Threshold of War

The first case study explores the 2018–2022 protests in Basra, Iraq, as an illustration of how climate change can contribute to conflict below the threshold of war. Over the past five years, Basra has been the site of significant protests that often coincide with the hottest temperatures of the summer.[2] The protests began in response to conditions that are closely related to extreme heat and water scarcity but grew into a larger movement with demands that extended beyond those linked to climate stressors.[3] Initially, insufficient electricity for refrigeration and cooling, which was exacerbated by extreme heat, and a lack of clean water suitable for drinking and agriculture, which was driven by environmental mismanagement and exacerbated by dryer conditions, fueled protests over poor public services. Those grievances ultimately interacted with broader governance concerns, building into a national movement that eventually unseated the government of Prime Minister Adil Abdul-Mahdi. Even after the change in national leadership, protests in Basra and surrounding governorates continued.[4]

Climate Hazards Exacerbate Poor Public Services

Southern Iraq is particularly prone to extreme heat associated with global warming. Figure 3.1 shows the incidence of extreme heat relative to a historic baseline in Iraq's three southernmost governorates—Basra, Dhi Qar, and Maysan—which straddle where the Tigris and Euphrates rivers meet. The outbreak of protests in 2018 coincided with a period that included roughly 120 days of temperatures at or exceeding 110°F, which was substantially higher than the historic baseline of extreme heat. The effect of the extreme heat intersected with electricity shortages, magnifying the environmental challenge since electricity was critical to coping strategies: air conditioning for relief from the heat and refrigeration to prevent food spoilage.[5]

Figure 3.1. Map of Iraq's Southern Governorates and Incidence of Extreme Heat

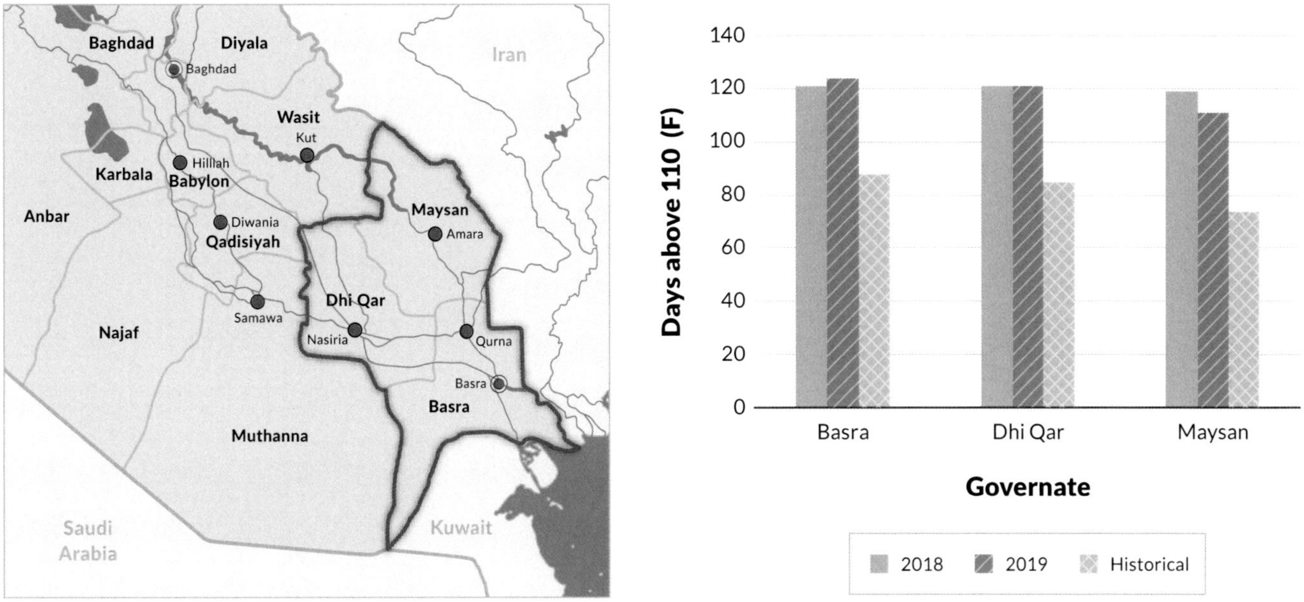

SOURCE: Features temperature data from the Copernicus archive accessed from Berkeley Earth, database, undated. The data reflect global surface temperatures from Berkeley Earth Surface Temperatures (BEST).

The impact of extreme heat was compounded by a second environmental challenge facing the southern governorates: a lack of clean water. This environmental challenge stemmed from a variety of causes, including the former regime's draining of the southern marshes, the lack of maintenance on water pumps and water treatment plants, the damming of tributaries in both Iraq and Iran, and the lack of information and urgency in responding to the initial outbreak of contaminated water.[6]

Extreme heat in Iraq has a direct negative impact on human health and agricultural yields. As it relates to human health, extreme heat interacts with humidity to drive wet bulb temperatures to the point at which the body cannot effectively cool itself, leading to heat stroke. Extreme heat also increases demand for electricity; when that demand goes unmet, it can drive grievances over limited access to air conditioning, food refrigeration, and clean water that relies on electric pumps or electric purification. Figure 3.2 depicts unmet peak demand in federal Iraq that coincides with the protest movement, with the orange bars representing shortfalls in electricity. Although there are no data available that are Basra-specific, there are reasons to think the shortfalls were even more severe in this area. First, Basra is one of the hotter locations in Iraq, which drives high electricity demand.[7] Second, both local reporting and regional reporting suggest that electricity blackouts in major Iraqi cities persisted for even longer periods in the day than reflected by the orange bars in Figure 3.2. Reporting commonly cited blackouts up to 20 hours daily.[8] One Basra resident summarized the dilemma as follows: "When it comes to electricity, it's even worse. Without air conditioning, it's hard to breathe. The government promised that we would have 20 hours of electricity per day this summer [but] we only have between four and six hours of electricity per day."[9] The resident's comments are instructive in that, in addition to identifying the hardship—only four to six hours of electricity per day—the statement emphasizes the government's inability to meet its commitments. The electricity strain not only prevented cooling, it also contributed to poor quality drinking water that sickened individuals. In summer 2018, there were more than 100,000 cases of illness from water contamination in Basra governorate alone.[10] The water was even unsuitable for agriculture because of high salinity levels. The problem of salinity results from an increase in water usage upstream combined with evaporation. This situation allows salt water from the Persian Gulf to further encroach into the Shatt Al-Arab River and its canals, which support agricultural irrigation in Basra.[11] Although 2018 marked a spike in illness from this environmental problem, the effect on livelihoods had been building. Cultivated land in Basra governorate had been cut in half in the ten years prior to 2018.[12] These overlapping crises—extreme heat and water scarcity—were at the heart of the grievances that drove Basra's initial protests.

Figure 3.2. Unmet Peak Electricity Demand in Federal Iraq

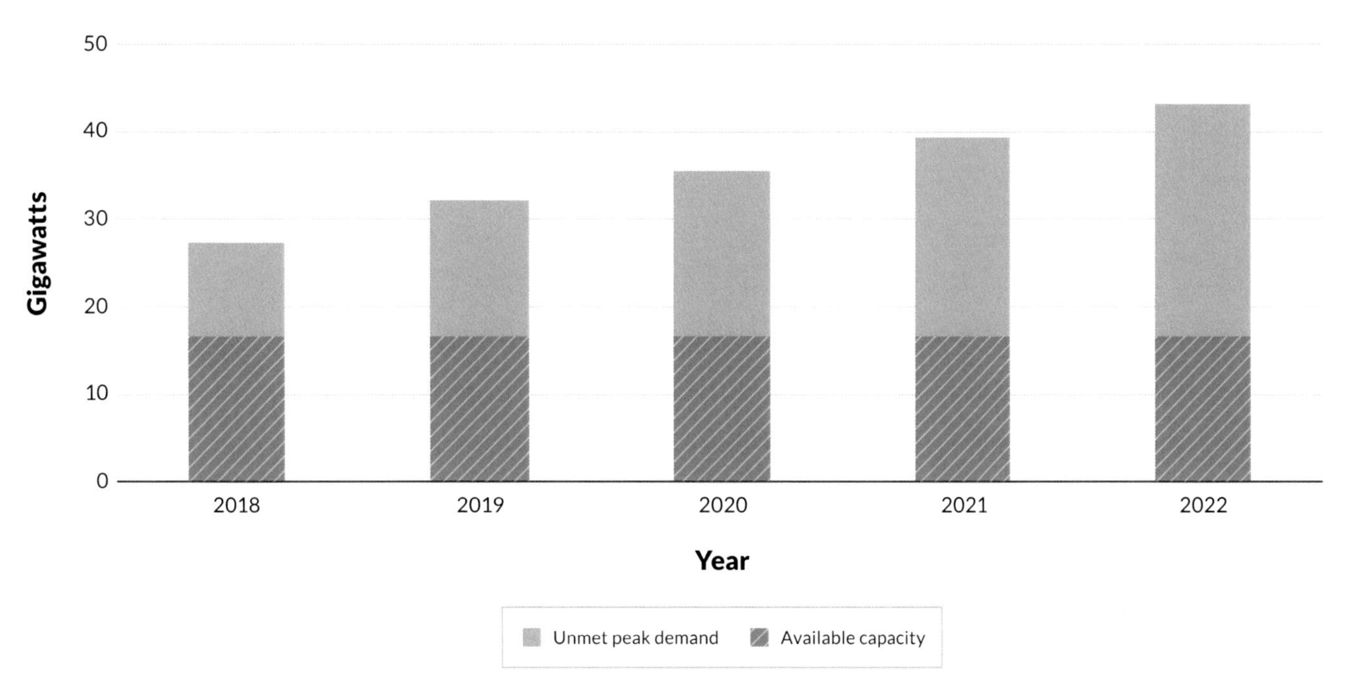

SOURCE: Features International Energy Agency projections with Iraq's electricity production held constant from 2018 against forecast peak demand. (International Energy Agency, "Iraq's Electricity Supply and Demand, 2018–2030," webpage, last updated April 25, 2019).

Resulting Conflict

The protests in Basra and surrounding governorates began in early July 2018, when daily high temperatures regularly exceeded 118°F. The protests coincided with Iran cutting electricity exports to Iraq, which limited citizens' coping strategies.[13] Journalistic interviews with protestors consistently revealed three primary grievances: water shortages and deteriorating water quality, power cuts, and lack of employment opportunities.[14] Juxtaposed against these grievances was Basra's pivotal role in Iraq's energy sector; the province produces the overwhelming majority of Iraq's crude oil. Not coincidentally, some of the earliest protests were staged at oil facilities in Qurna, Burjisiya, and Rumaili, which gave symbolic importance to protestors' demands for public services commensurate with the area's role in generating Iraq's oil wealth.[15]

As shown in Figure 3.3, the protests were largely peaceful, but there were violent incidents. Protestors blocked roads, set fires, destroyed property, and clashed with security forces.[16] Security forces also responded with violence. In the first three months of the protests, an estimated 27 Iraqis were killed in the unrest, with many more injured.[17] In one of the deadliest weeks of the protests in September 2018, protestors stormed and set fire to Basra provincial government buildings and the Iranian consulate.[18] Iraqi officials, including then-Prime Minister Haider al-Abadi, visited Basra with promises to address the protestors' demands. Grand Ayatollah Ali al-Sistani expressed solidarity with the protestors, acknowledging that "many of [Basra's] people suffer from hardship, few public services, the spread of diseases and epidemics, and the majority of [Basra's] youth do not find work opportunities commensurate with their capabilities and qualifications."[19] As shown in Figure 3.3, protests continued in Iraq's southern governorates beyond 2018, and there has been a clustering of protest events during the summer, which has become known as Iraq's "season of protests."[20] However, what is different about the post-2018 iterations of these protests is the stronger link to national-level political issues. In 2019, Basra's summer protests ultimately merged into the October Revolution, which drew in the majority of Iraq's governorates and eventually led to the resignation of the Adil Abdul-Mahdi government. More recently, the protests have intersected with Muqtada al-Sadr's political aspirations, debates inside Iraq over foreign influence, and shocks to the Iraqi political system, such as the U.S. decision to kill Iranian Islamic Revolutionary Guard Corps-Qods Force commander Qassem Soleimani and a high-ranking Iraqi leader of the Popular Mobilization Units on Iraqi territory.[21]

Causal Pathway

Figure 3.4 depicts the steps that played out within the general causal pathway for intrastate conflict introduced in Chapter 1. In the initial steps after the climate hazards, the public health crisis (i.e., illness from contaminated water) and the further degradation of livelihoods via the decline in arable land drove the most important impacts. Iraq's economic crisis—although not derived from climate change—can be considered an accelerant of the evolution of the conflict. The economic crisis denied the state the ability to import electricity that would have provided key coping mechanisms for the local population. Instead, Iraq fell behind on its estimated $100 million monthly bill to Iran for electricity, prompting Iran to cut off supply.[22] And Baghdad's fiscal challenges reinforced preexisting grievances around corruption and inequality that the climate hazards exacerbated.

As the crisis moved toward conflict, a lack of government legitimacy emerged as perhaps the most important factor in the potential for a violent escalation. Poor delivery of government services was a central grievance of protestors. Once the protests devolved into the storming of government facilities, arson, and road blocks, an inept government response, which included the excessive use of force against demonstrators, created a negative feedback loop that further drove the use of violence by demonstrators.[23]

A limitation of the causal pathway, however, is that it does not illuminate how climate hazards interacted with broader grievances in Iraq. Although climate hazards provided a spark for the movement, protestors in Basra raised issues that run deeper than environmental concerns, varying from the role of Iran in Iraqi decisionmaking, to the role of informal ethno-sectarian quotas in Iraqi governance, to the corruption of the state's political class.[24] In a particularly stark assessment of governance quality, an Iraqi political scientist lamented, "The post-2003 governments were not able to meet even the minimum citizen rights. The competing parties were preoccupied with politics, representation in government, quarrelling and dealmaking. They considered Iraq loot from which to take shares."[25]

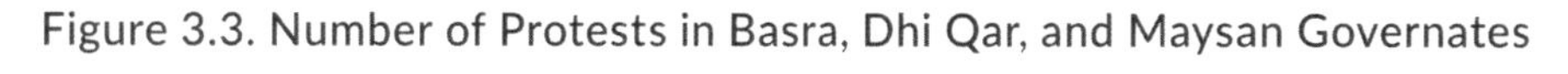
Figure 3.3. Number of Protests in Basra, Dhi Qar, and Maysan Governates

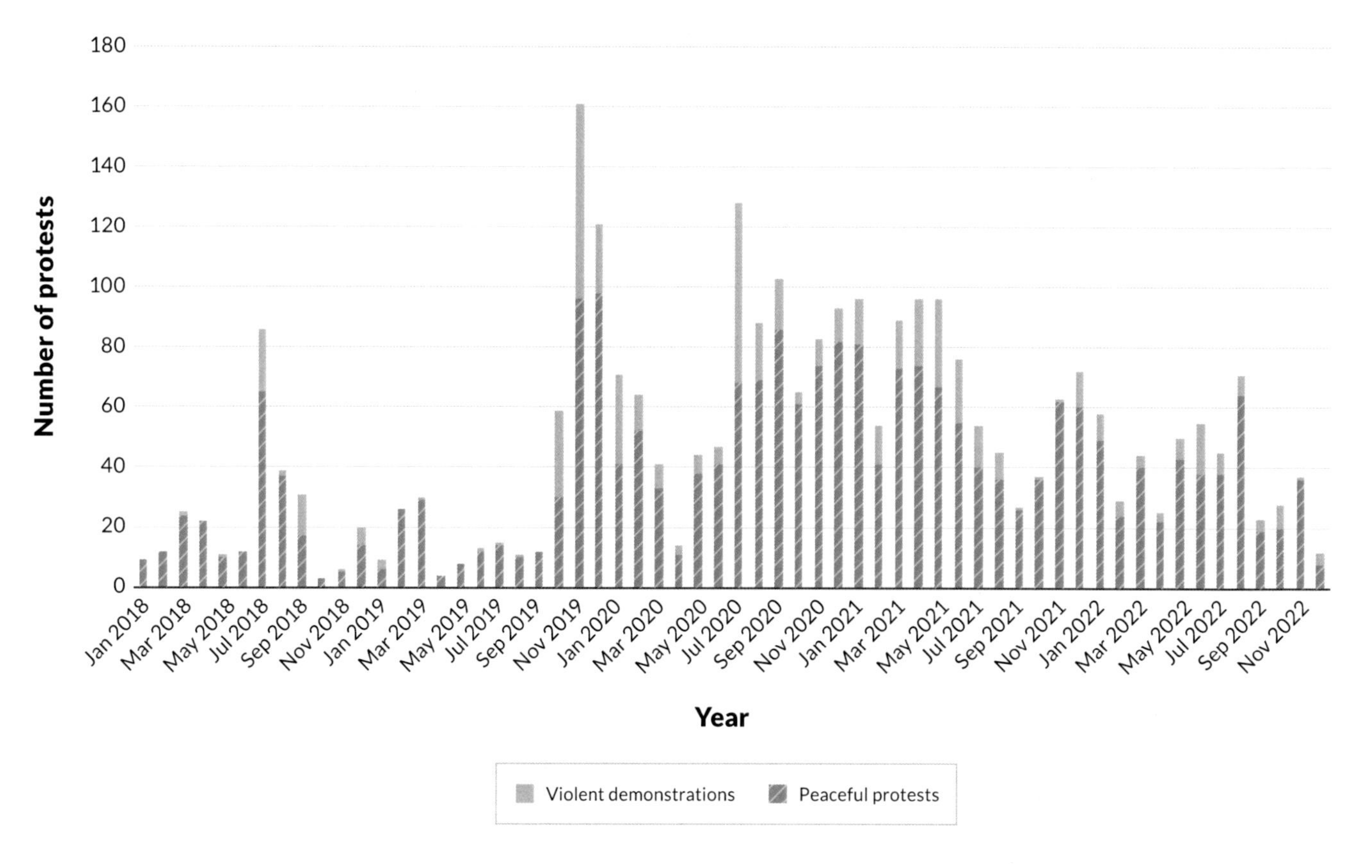

SOURCE: Features data from the Armed Conflict Location and Event Data Project (ACLED), "Data Export Tool," database, last updated June 2023.

Figure 3.4. Causal Map from Climate Change to Violent Protests in Southern Iraq

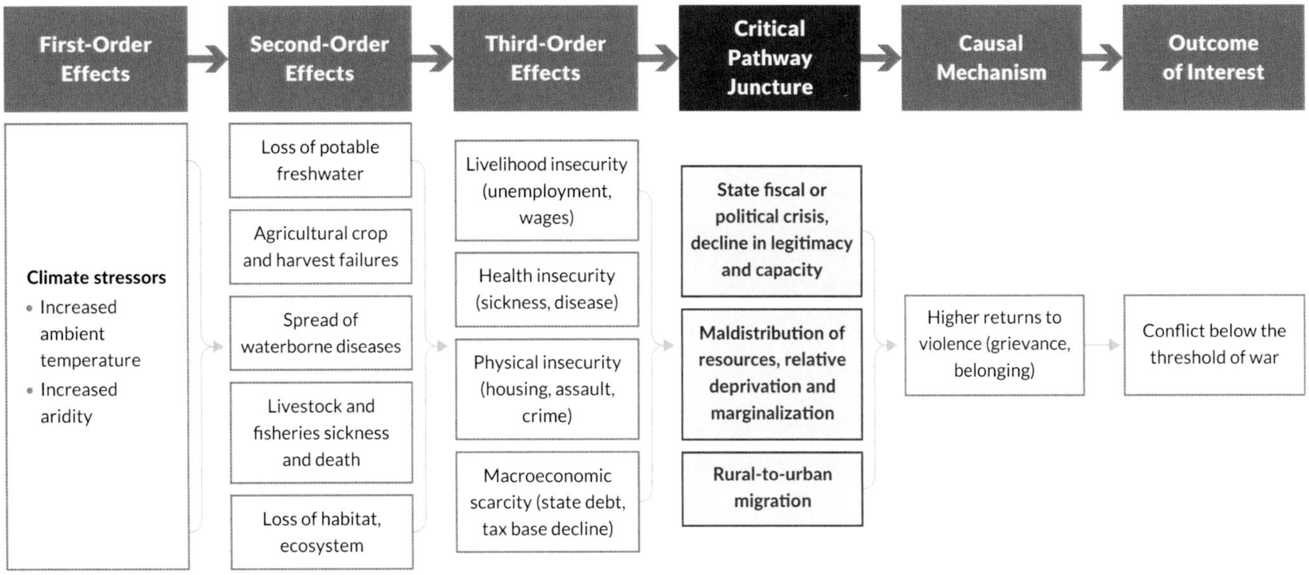

First-Order Effects → **Second-Order Effects** → **Third-Order Effects** → **Critical Pathway Juncture** → **Causal Mechanism** → **Outcome of Interest**

First-Order Effects

Climate stressors
- Increased ambient temperature
- Increased aridity

Second-Order Effects

- Loss of potable freshwater
- Agricultural crop and harvest failures
- Spread of waterborne diseases
- Livestock and fisheries sickness and death
- Loss of habitat, ecosystem

Third-Order Effects

- Livelihood insecurity (unemployment, wages)
- Health insecurity (sickness, disease)
- Physical insecurity (housing, assault, crime)
- Macroeconomic scarcity (state debt, tax base decline)

Critical Pathway Juncture

- **State fiscal or political crisis, decline in legitimacy and capacity**
- **Maldistribution of resources, relative deprivation and marginalization**
- **Rural-to-urban migration**

Causal Mechanism

Higher returns to violence (grievance, belonging)

Outcome of Interest

Conflict below the threshold of war

Basra Conclusion

This short case study, which profiled violent protests in Iraq that erupted during a period of extreme heat and water scarcity, demonstrates that climate change can be a contributing factor to conflict below the threshold of war in the CENTCOM AOR. Specifically, the Basra case study suggests that environmental factors can interact with other grievances, including national-level politics, to fan a political movement. Actors on both sides of the movement—those pushing for and those resisting change—employed violence to advance their interests, and that conflict raised the risk of a broader war that could have precipitated a U.S. intervention.

The 1970 Bhola Cyclone as an Illustration of Intrastate War

The second case study examines how an extreme climate event can contribute to the onset of intrastate war. Using the Bhola cyclone that devastated East Pakistan in 1970 as the backdrop, our research explores how that event exacerbated preexisting grievances that culminated in a civil war the following year. The analysis does not necessarily attribute the Bhola cyclone to climate change, as the existence of cyclones in South Asia predates the Industrial Revolution, which climate change is commonly associated with. And irrespective of the Bhola cyclone, the subsequent conflict would have transpired. We argue the cyclone accelerated existing, underlying grievances between East and West Pakistan that would have eventually escalated into intrastate war. Our research design explores how an extreme weather event—an occurrence that is forecast to become more frequent and more intense throughout the 21st century—can interact with other existing factors to precipitate conflict. In this way, the Bhola cyclone serves as a proxy to explore a future CENTCOM AOR beset by such extreme weather events.

On November 12, 1970, the Bhola cyclone made landfall in East Pakistan (present-day Bangladesh). There is debate over whether the danger was communicated to the local population, with some sources claiming that up to 90 percent of the population received warning of the cyclone, although others claim that authorities failed to provide sufficient warning despite information provided to Pakistani authorities by the U.S. National Environmental Satellite, Data, and Information Service, part of the National Oceanic and Atmospheric Administration.[26] Whatever the truth, few people appear to have evacuated the area at risk. High winds, flooding, and a lack of evacuation and other mitigation measures combined to make the cyclone one of the deadliest natural disasters of the 20th century, affecting more than 4 million people and killing at least 250,000.[27]

The apathy and incompetence displayed by the Martial Law Administration while conducting relief efforts in the aftermath of the cyclone would solidify the Bengali population's alienation from the state apparatus of Pakistan, contributing to a landslide electoral victory for the East Pakistan Awami League in elections held just a month later in December 1970. The political crisis that ensued resulted in a crackdown by the Martial Law Administration on the Awami League and Bengali nationalists, sparking a civil war that drew in India. That civil war is estimated to have killed between 1.7 million and 3 million people and sent about 10 million displaced persons to neighboring India, although the true number is unknown because of poor recordkeeping.[28] In this case study, we examine the role played by the cyclone and its aftermath in accelerating the events that led to this conflict.

Catastrophic Climate Event Exacerbates Long-Standing Tensions

The cyclone and the poor state relief response underscored the marginalization of East Pakistan by Pakistani decisionmakers in Islamabad. Tensions between the two wings existed as early as the 1947 independence of Pakistan and continued to fester until reaching a boiling point two decades later. When Pakistan became an independent country after the partition of the subcontinent, the bulk of its political, bureaucratic, and military leadership consisted of individuals from West Pakistan (present-day Pakistan) or migrants with roots in territories that became part of India. Independent Pakistan found itself in a weak position. The country lacked contiguous territory connecting its two parts, was economically underdeveloped, was filled with diverse ethno-linguistic groups whose only unifying characteristic was religion, and was faced with hostile India and Afghanistan as neighbors (see Figure 3.5 for a geographical depiction of East and West Pakistan).[29] To compensate, the founders of Pakistan took measures to create a strong central government and attempted to unify the country by establishing Islam as the state religion and Urdu as the single national language. They also cracked down on opposition elements, such as communists, and invested in significant resources to strengthen the armed forces.[30] These measures were predictably unpopular in East Pakistan, where the most prevalent language was Bengali and a significant part of the population was non-Muslim. East Pakistan also had minimal representation in the state apparatus, such as the civil service and the military.[31] In early 1948, riots in East Pakistan demanding the elevation of Bengali as a national language were violently suppressed. Despite concessions establishing Bengali as a state language in East Pakistan, the damage was already done.[32]

These differences between East and West Pakistan would deepen in the following decades. In 1954, the government of Pakistan established the One Unit Scheme (discarded in March 1970), whereby East and West Pakistan would have equal representation in parliament. Since East Pakistan had a larger population, this measure effectively diluted its political voice. East Pakistan was also deprioritized in defense planning; despite Pakistan receiving significant military aid in the 1950s and 1960s, from which it built up large and well-equipped armed forces, its defense planners did not allocate significant forces to East Pakistan's defense.[33] Moreover, East Pakistan harbored economic grievances against the western wing. After a successful military coup led by General Muhammad Ayub Khan in 1958, Pakistan witnessed a Decade of Development from 1958 to 1969.[34] The combination of political stability and significant aid from the United States and international institutions allowed Pakistan to achieve an annual average growth rate of 6.7 percent from 1960 to 1970 while reducing poverty. However, much of this economic windfall was reaped in West Pakistan, as indicated by the large gap in the country's per capita gross national product: 504 rupees in West Pakistan and 314 rupees in East Pakistan (or $105.90 and $66.00 in 1970 U.S. dollars).[35] This disparity was especially galling since a significant portion of Pakistan's foreign exchange was earned from the jute industry based in East Pakistan.[36] East Pakistan was also deprioritized in development planning. In 1956, East Pakistan was allocated only 3 billion rupees ($630 million) from a total development budget of 9.23 billion rupees ($1.94 billion).[37]

When considering these factors, it is evident that the Bengali population of East Pakistan felt alienated from West Pakistan. However, the poor planning and apathy displayed by Pakistani authorities in their response to the Bhola cyclone likely exacerbated these grievances. Coastal Bangladesh experiences cyclones on a near-annual basis, with the bulk of these events occurring within two windows: premonsoon (April to May) and postmonsoon (October to November). The history of cyclones in the Bengal region indicates that premonsoon cyclones are more common, but the most devastating cyclones, *super-cyclones*, tend to occur in the postmonsoon season.[38] When a cyclone makes landfall, coastal areas are inundated by water in a phenomenon known as *storm surges*, which are responsible for most of the destruction associated with cyclones.[39] The low-lying and densely populated coastal areas of Bangladesh are particularly vulnerable to cyclones and storm surges, with each storm bringing the potential to cause immense devastation. According to one estimate, up to 42 percent of global deaths from cyclones in the past two centuries occurred in what is now southern Bangladesh.[40]

Figure 3.5. West (Green) and East Pakistan (Orange), 1947–1971

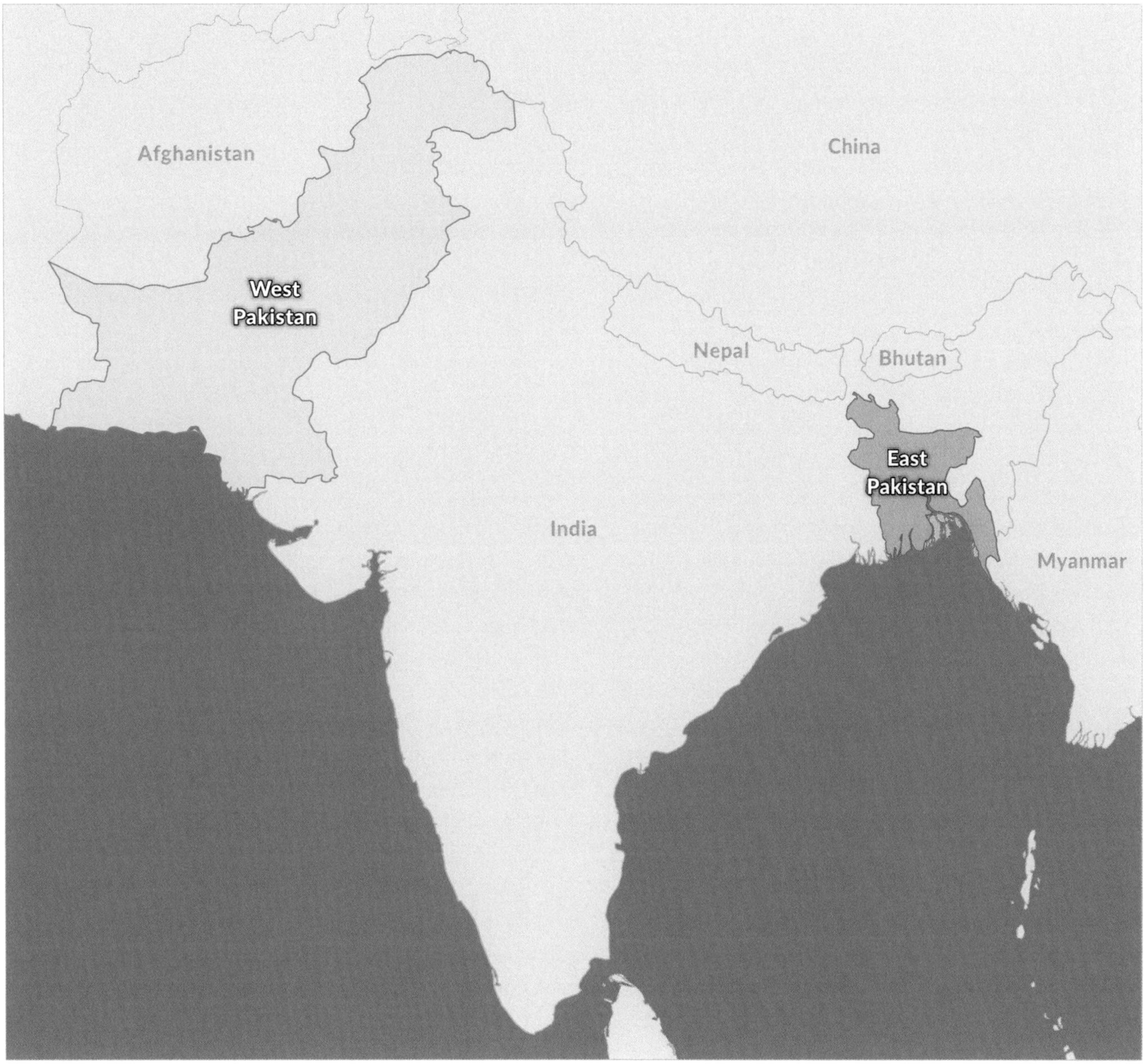

From 1959 to 1969, East Pakistan witnessed an average of almost two cyclones per year. In the wake of 20,000 deaths during the 1960 cyclone season, the government of Pakistan acknowledged the inadequacy of its disaster mitigation strategy and consulted with the U.S. National Hurricane Center. The main recommendations from this consultation included the establishment of embankments and cyclone shelters along the coastline.[41] The government also established a hierarchy to manage relief efforts during disasters; in July 1970, official guidelines were issued that emphasized the role of the Pakistan Meteorological Department's warning system in mitigating cyclone impacts. Few of these measures appear to have been implemented seriously, however. When the Bhola cyclone hit, many of the affected zones did not receive warnings because of the lack of radios or communications infrastructure; among those who did receive warnings, many refused to leave their property because of fear of theft or the lack of appropriate shelters.[42] Commenting on the negligence of the Pakistani authorities, an expert from the U.S. National Hurricane Center claimed that up to 90 percent of deaths could have been avoided if even basic shelters had been erected along the coastline.[43] The Bhola cyclone also devasted farming fields, immediately resulting in food insecurity for millions of Bengalis who lived "meagre lives" from the soil in southern East Pakistan.[44]

The negligence from Pakistani authorities to mitigate the impact of the cyclone was compounded by their slow response. It took several days for news of the disaster to arrive in Dhaka; when it did, East Pakistan political leaders requested that the authorities declare a state of emergency—a request that was refused. In some districts of East Pakistan, aid workers reported that there were no government relief efforts even ten or 12 days after the cyclone hit. Although people in West Pakistan donated generously to relief efforts, poor planning limited the amount of essentials that were distributed, and the authorities refused to transfer helicopters to East Pakistan to assist with relief operations.[45] The sluggish start was exacerbated by the optics, with President Yahya Khan making only a cursory aerial survey of some affected areas. Other countries, such as Iran, responded more effectively than the Pakistani government itself, with Iran declaring a national day of mourning and sending extensive aid. The top-down approach of Pakistani authorities hampered the operations of foreign aid workers, which prompted some organizations to sever relations with the government and operate independently because of the poor coordination.[46]

Although the government's role in the relief efforts eventually improved, the cyclone further diminished the legitimacy of the government of Pakistan for the eastern wing. When Ayub Khan resigned as president in 1969, he was replaced by then-General Yahya Khan as president and chief martial law administrator. Khan planned to revert to democracy and announced that elections would be held in 1970. Crucially, he discarded the One Unit Scheme and announced that voting would be conducted through universal adult franchise, which would entitle East Pakistan to a larger proportion of the seats in the National Assembly because of its larger population. The Awami League of Sheikh Mujib-ur-Rahman (known as Mujib) was the dominant party in East Pakistan. Mujib unveiled a policy platform titled the Six Points; according to this manifesto, the eastern and western wings of Pakistan would be autonomous, maintaining separate currencies, fiscal accounts, economic policies, and armed forces, with the central government only having authority over defense and foreign policy.[47] Buoyed by a recent stint in prison as a political prisoner, the apathetic government response to the cyclone, and an electoral boycott of East Pakistan's other major party led by influential Bengali politician Maulana Bhashani, Mujib led the Awami League to a landslide victory in December 1970, securing 160 of 162 National Assembly seats assigned to East Pakistan and attaining a simple 300-member majority in the national legislature. In West Pakistan, Zulfiqar Ali Bhutto's Pakistan People's Party (PPP) won 81 of 138 seats.

Resulting Conflict

Although the Awami League had a clear mandate, the election results sparked a political crisis. Bhutto and many of the West Pakistani elite were reluctant to hand over power to Mujib, since his Six Points would effectively make both wings of Pakistan separate states with only a nominally unified structure. Bhutto refused to attend the inaugural session of the new National Assembly and prevented any of his party's representatives from attending, resulting in the postponement of the session. President Yahya Khan attempted to mediate between Mujib and Bhutto, but neither was willing to budge. Sensing that their electoral victory was being repudiated, the Awami League put out a call on March 7, 1971 for civil disobedience in East Pakistan. The situation rapidly deteriorated, with clashes erupting between the Pakistani military and Awami League activists and attacks by Bengali nationalists on non-Bengali populations. The government began moving in reinforcements from West Pakistan. On March 25, Pakistani authorities launched a crackdown called Operation Searchlight with the aim of arresting Mujib and the Awami League leadership, disarming the majority Bengali East Pakistan Rifles paramilitary units, and suppressing the opposition in East Pakistan.[48]

Although the government did arrest Mujib and some other senior leaders, most of the Awami League's lower-level leadership evaded arrest. Likewise, the efforts to disarm the East Pakistan Rifles were only partially successful, and many Bengali personnel mutinied independently. By April 10, Pakistani forces controlled various major cities and the vital military cantonments and airfields, but the rest of East Pakistan was outside government control. Over the following two months, the Pakistani Army and other paramilitary units had to contend with fierce guerilla warfare as they attempted to reestablish control over all of East Pakistan. By June 1971, buoyed by reinforcements and rebels who were hampered by a lack of equipment and training, the Pakistani army managed to drive the remnants of the Bengali resistance into India. However, this did not end the conflict; millions of Bengalis had fled the fighting for India, and a government in exile formed there on April 17.[49] The Indian government provided training, munitions, and logistical support to the Bengali resistance, which organized under the banner of the Mukti Bahini, a resistance group of Bengali military personnel and civilians. Starting in June, the Mukti Bahini began to launch irregular attacks, first from bases in India but increasingly from within East Pakistan, that forced the Pakistani military to withdraw from indefensible positions. Although Pakistani forces successfully contained the Mukti Bahini's monsoon offensive in summer 1971, by the end of the year, the situation deteriorated into a virtual state of war between India and Pakistan, with Indian forces and the Mukti Bahini launching joint operations by November.[50]

Believing that East Pakistan was indefensible in the long run with India supporting the Mukti Bahini, Pakistan's leadership determined that a rapid offensive against India in the west would compel negotiations and result in a ceasefire in the east. To establish air dominance, Pakistan's air force launched preemptive strikes on Indian airfields along West Pakistan's border the night of December 3. However, these strikes inflicted only minimal damage and gave license to India to openly join the war effort in the east, where it had been building up forces since April. Over the following two weeks, India's armed forces blunted Pakistan's offensive in the west while launching their own offensive in the east, culminating in the fall of Dhaka on December 16, 1971. That same day, Lieutenant General Amir Abdullah Khan Niazi, the commander of Pakistani forces in East Pakistan, signed the Pakistani Instrument of Surrender.[51]

Causal Pathway

Figure 3.6 depicts the steps that played out within the general causal pathway introduced in Chapter 1. In the initial steps after the Bhola cyclone, the sheer scale of the disaster, the Pakistani government's apathetic emergency response, the displacement of millions of East Pakistanis, the resulting public health crisis (e.g., illness from contaminated water, malnutrition), and the further degradation of livelihoods via the decline in arable land drove the most important impacts. These factors, in turn, increased competition for already scarce resources, which were unevenly distributed in the cyclone's aftermath.

Long-standing Bengali resentment toward West Pakistan and West Pakistani indifference toward its eastern province—with neither viewpoint derived from climate change—can be considered accelerants for the tinder that flamed into the 1971 war. Instead of quickly aiding East Pakistan after the Bhola cyclone, West Pakistani officials dithered, worsening the conditions of millions of already impoverished people and reinforcing widespread and long-standing sentiments that East Pakistanis were second class citizens. When West Pakistani officials finally intervened, their actions proved to be inept and insufficient, creating a negative feedback loop that only emboldened East Pakistanis to shift their allegiance to the Awami League, if they had not already done so.

As tensions escalated, the PPP's refusal to recognize the Awami League's success in the 1970 general election emerged as potentially the most important factor in the prospect for violent confrontation. Protestors who first galvanized around shared sentiments of West Pakistani apathy and ineptitude in the delivery of basic services following the Bhola cyclone now protested over the PPP's refusal to hand over power. This led to the Awami League's call for civil disobedience, a move that rapidly deteriorated into violence with West Pakistani forces and, ultimately, Operation Searchlight and the beginning of the 1971 civil war. Mass migration ensued; at least 10 million Bengalis fled to neighboring India, which provided support to Bengali rebels. India also repelled a West Pakistani military effort on its western front that was intended to distract from that support.

As in the Basra case, the climate event we consider here is a factor in the resulting conflict, interacting with much deeper preexisting grievances rather than serving as the main driver of the 1971 war. The climate event contributed to the further delegitimization of West Pakistan's political elite that led to an electoral rout by a political movement in East Pakistan bent on achieving autonomy, if not separation. This led to civil war, in which Pakistan's more powerful neighbor—India—had a major stake in further weakening its rival.

East Pakistan Conclusion

This case study illustrates that an extreme climate event can contribute to the onset of civil war and that the phenomenon has already occurred in the CENTCOM AOR. However, the analysis also shows that the relationship between extreme weather events and high-intensity conflict may be indirect and that the extreme weather event may only serve as a contributing factor to conflict onset when conditions are already ripe for the outbreak of war. Although the Bhola cyclone may not have been caused by climate change, it is instructive for exploring the relationship between extreme weather events and conflict in the CENTCOM AOR. This is because climate scientists project that the severity and frequency of cyclones will increase in the future and that the rate of change will be faster in the Arabian Sea, which Pakistan borders. This change will increase the risk of the type of climate-related conflict explored in this case, even if extreme weather events are just one factor in a broader array of conditions that result in civil war onset.

Figure 3.6. Causal Map from 1970 Bhola Cyclone to 1971 Civil War

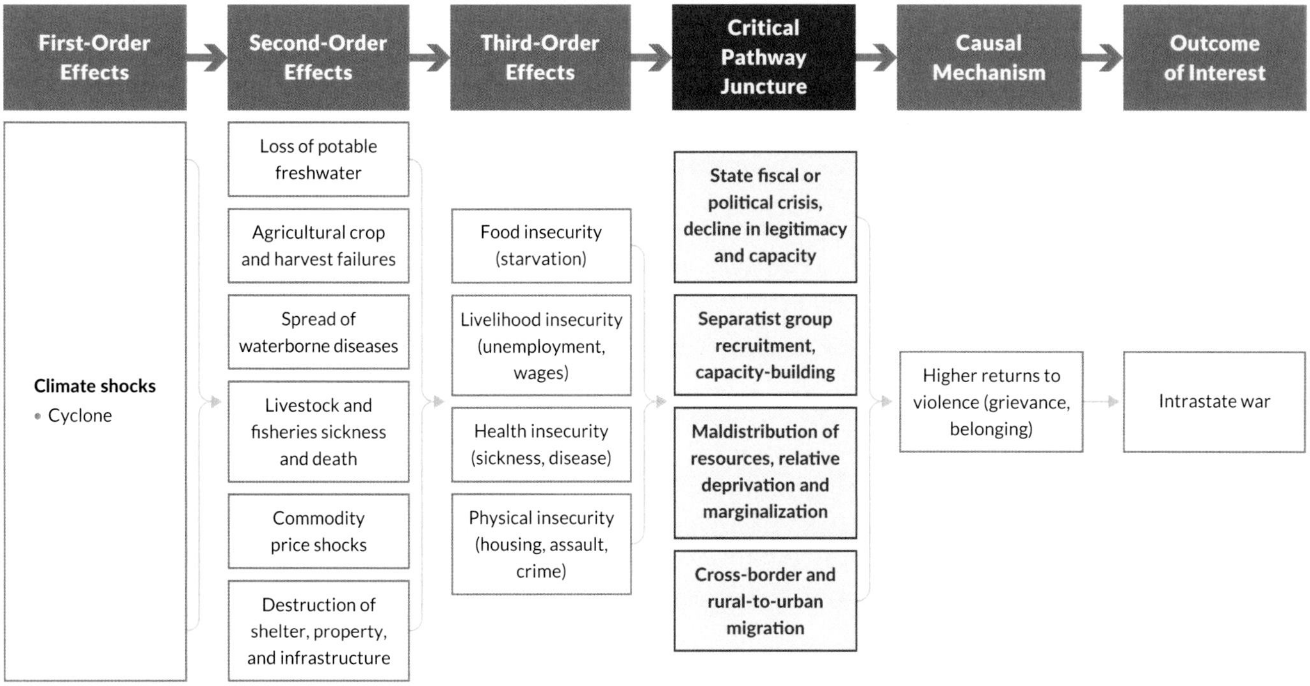

First-Order Effects	Second-Order Effects	Third-Order Effects	Critical Pathway Juncture	Causal Mechanism	Outcome of Interest
Climate shocks • Cyclone	Loss of potable freshwater Agricultural crop and harvest failures Spread of waterborne diseases Livestock and fisheries sickness and death Commodity price shocks Destruction of shelter, property, and infrastructure	Food insecurity (starvation) Livelihood insecurity (unemployment, wages) Health insecurity (sickness, disease) Physical insecurity (housing, assault, crime)	**State fiscal or political crisis, decline in legitimacy and capacity** **Separatist group recruitment, capacity-building** **Maldistribution of resources, relative deprivation and marginalization** **Cross-border and rural-to-urban migration**	Higher returns to violence (grievance, belonging)	Intrastate war

The Grand Ethiopian Renaissance Dam as a Potential Flashpoint for Future Interstate War

The third case study examines how climate change may contribute to the onset of the rarest form of conflict: interstate war. We are not aware of a prior interstate war in the CENTCOM AOR in which environmental factors were a primary contributor. So, unlike the first two case studies, which profiled historical episodes, this case study focuses on the risk of a future interstate war between Egypt and Ethiopia over the GERD. To be clear, the environmental stress and resource competition that may result from the GERD is not a climate issue per se; it is about the downstream effects of large-scale environmental management. However, the anxieties of the Nile River's lower riparian states about the GERD are compounded by climate change, since the feared impact of the GERD—water scarcity and land degradation—will also occur via climate change. Our research design examines the risk of a future interstate war between Ethiopia and Egypt—the two most powerful Nile River riparian states—over the consequences of the GERD. To evaluate the seriousness of that risk, we use recent defense acquisitions by these two states and Egypt's defense relationship with Sudan as indicators of whether these countries are preparing for the possibility of war. We find that they are preparing for this contingency, validating the proposition that climate change interacting with other factors could contribute to the onset of interstate war along this seam of the CENTCOM AOR.[52]

The GERD has been under construction since 2011, and Ethiopia, Sudan, and Egypt have been involved in fractious negotiations to regulate its operation for nearly as long. The three countries have engaged various outside actors, such as the African Union, the United States, and the World Bank, in the negotiations.[53] Although the process has allowed for progress on some matters of interest, the most contentious matters—related to the duration of the dam's filling period and its operation—remain unresolved.[54] Disagreement and tensions have repeatedly stalled negotiations, with Egypt threatening military action at points.[55] Tensions escalated most recently in 2020 after Ethiopia decided unilaterally to begin filling the dam, despite the lack of a binding agreement. Ethiopia remains against the type of legally binding agreement that would satisfy Sudan and Egypt.[56] In light of this, Egypt expressed its view that the dam is an existential threat, making the prospect of a resolution that satisfies all countries uncertain.[57]

Both Ethiopia and Egypt have made military threats over the GERD, although such threats likely are part of a larger negotiating strategy. In 2021, Egyptian President Abdel Fattah

El-Sisi sent Ethopia a warning: " [L]et's not reach the point where you touch a drop of Egypt's water, because all options are open."[58] In March 2023, Sameh Shoukry, Egypt's foreign minister, reemphasized this position, stating that "all options are open and all alternatives remain available."[59] Tensions are highest when the GERD reservoir is filling, and it is during those filling times when Ethiopia has emphasized its military capabilities—including the development of its air force—to repel any attacks. For example, in the context of the July 2022 filling, Ethiopia announced heightened military alert levels and vowed that "our Air Force is in a position to protect and defend our airspace from any kind of attack."[60]

Climate Change Exacerbates Environmental Management Challenges

Ethiopia has struggled to provide electricity to its population; only about 50 percent of Ethiopians have access to electricity, and such places as primary schools and health clinics are estimated to have even lower levels of access.[61] In response, the Ethiopian government launched the GERD project—a hydroelectric dam on the Blue Nile aimed at doubling Ethiopia's electricity output.[62] The dam has the potential to support economic development for Ethiopia, including enhancing agricultural production, and can offer some of these benefits to Sudan, particularly in terms of irrigation, flood reduction, and electricity.[63] Additionally, the dam may be able to help regulate Nile water flow in Egypt during times of water scarcity.[64]

Egypt, downstream of Ethiopia and heavily reliant on the Nile, already deals with water scarcity that climate change, poor water management, and population growth continues to exacerbate.[65] Additionally, "although 85% of Nile waters originate in Ethiopia [with the Blue Nile], nearly all consumptive use occurs downstream in Egypt and Sudan."[66] Most of Egypt's water—around 93 percent (of which 85 percent is used for agriculture)—comes from the Nile.[67] The Blue Nile's water comes from "the highly variable monsoon-driven rain from the Ethiopian highlands."[68] One study found that rainy seasons over the Blue Nile Basin are producing significantly less precipitation and that droughts are increasing in frequency.[69] Even if precipitation increases as a result of climate change, hot and dry years, which are also increasing in frequency, have affected Nile River flows and food security in Egypt and will likely continue to do so.[70] These inconsistent conditions can cause agricultural price fluctuations, which can exacerbate existing food insecurity and have the potential to increase agricultural unemployment.[71] Additionally, rising sea levels near the Nile Delta combined with wastewater recycling techniques—intended to mitigate the effects of water scarcity on Egypt's ability to meet growing agricultural demands—have increased Nile water salinity.[72] As it stands, "water scarcity and salt water intrusion, compounded by other climate change indicators, threaten to have a catastrophic impact on Egypt's agricultural output, soil quality, and water supply."[73]

Changes to the Blue Nile river flow resulting from the GERD could intersect with existing water challenges and further exacerbate scarcity in Egypt. Research has already shown a more than 14 percent loss in the surface area of the Aswan Dam reservoir—Lake Nasser—likely as a result of Ethiopia filling the GERD.[74] This means that Egyptian water security is highly dependent on GERD reservoir operations and the upstream climate in Ethiopia. Decreased water availability because of the dam could "[render] a significant part of an otherwise fertile land less suitable for farming and other agricultural activities."[75] Quicker filling durations could also reduce water availability, particularly if a drought were to occur during the filling period without an Egyptian mitigation strategy; longer filling durations could lead to lower water losses that are spread out over a longer period if there is similarly no mitigation strategy.[76] However, "if managed in the correct manner, shifting water supply (storage) from the Egyptian desert to the Ethiopian Highlands may in fact increase Egypt's water supply over time, since Ethiopia has a much lower average temperature than Egypt, which implies that the evaporation rate of water will be slower in the Ethiopian reservoir than the reservoir in the Aswan Dam." Yet, if Sudan uses water from the dam for irrigation, these benefits could be offset.[77] Once the GERD is operational, the dam will not necessarily make the effects of droughts more severe. However, because of the sensitivity of the issue, the GERD may be perceived as having such an effect anyway, especially if no agreement governs "how reservoirs will be used and refilled post-drought."[78]

Negotiating a fill agreement, especially one that addresses filling during a potential drought, represents the largest concern for downstream countries, although concerns also exist regarding the long-term impacts of the dam on Nile River flow (this is especially salient with regard to future droughts).[79] Reaching an agreement regarding filing duration and drought management policies can mitigate these negative impacts for downstream countries, particularly Egypt. But negotiations between Ethiopia, Sudan, and Egypt have stagnated, and Ethiopia unilaterally finished the dam's third filling in 2022.[80] Ethiopia stands to benefit economically from a faster filling duration; however, this will likely cause more tension between the three countries. Additionally, "[the conflation of] Egypt's long-standing water scarcity challenges with exaggerated short-term impacts from filling the GERD reservoir" has likely increased the difficulty of reaching an agreement.[81]

Military Acquisitions

Although the probability of a future military confrontation over the GERD remains difficult to assess, some recent defense acquisitions by Ethiopia and Egypt appear to reflect preparation for this contingency. Specifically, Egypt's desire to acquire aircraft with greater combat ranges, build its aerial refueling capability, and acquire beyond-visual-range (BVR) air-to-air missiles are all consistent with this interpretation. Similarly, Ethiopia ringing the GERD with air defenses is almost certainly a counter to this potential threat. That said, it is not possible to narrowly tie all of Egypt's priority acquisitions to a GERD contingency. Egypt's investment in its air force capability is part of a much longer trend that spans different threat environments; in addition to the GERD, Egypt is likely considering potential contingencies in the Eastern Mediterranean or out-of-area operations with Arab Gulf partners.[82] Egypt's acquisitions may also be motivated in part by a desire to build relationships with other powers after the United States withheld some arms transfers in 2013.

That said, a process of elimination to explain Egypt's acquisitions suggests that a GERD contingency is an important consideration. Egypt's primary internal security threat is insurgency in the Sinai, which does not require advanced aircraft with longer range and/or BVR capabilities; Egypt would mainly need these capabilities to gain superiority in aerial combat or to conduct an aerial strike against a distant target.[83] Additionally, Egypt's relations with its historic rival Israel have moved from a cold peace to more strategic cooperation. Egypt and the Gulf states are exploring deeper security cooperation with Israel as a hedge against perceptions of waning U.S. commitment and the rising threat of Iran.[84] This leaves the two most likely external security threats that could motivate new air acquisitions: controlling the Mediterranean gas field in Egypt's exclusive economic zone (EEZ) and the GERD.[85] As assessed by two analysts who are familiar with the Egyptian military's thinking, "To effectively secure its EEZ, Egypt would need an effective aerial fleet with a fuel capacity larger than its short-legged F-16 fighters to support its operating naval units in the region."[86]

With respect to the GERD, if Egypt unilaterally, or possibly with Sudan, did resort to military action, it could conduct three possible types of operations: land, special forces, and air.[87] The most realistic way for Egypt to conduct a military operation against the GERD would be by air, as Egypt and Ethiopia do not share a border, and the large distance between the two countries significantly decreases the likelihood of a land operation.[88] To conduct an air operation, Egypt would likely desire BVR capability for its aircraft, to counter Ethiopia's BVR capabilities, and aircraft with longer combat ranges to more easily reach Ethiopian airspace.[89] As implied in the term, having BVR means that Egyptian fighter pilots would be able to engage Ethiopian aircraft beyond the range at which the pilots could physically see the opposing aircraft. This is a particularly important consideration for Egypt; although Egypt's air force is considered superior to Ethiopia's, Ethiopia already possesses BVR capability on its Su-27s, which means that Egypt would have to confront this disadvantage in air-to-air combat unless it also acquires BVR.[90] The refusal of the United States to transfer BVR capabilities to Egypt is a long-standing complaint of Cairo, with Washington withholding this capability on the basis of maintaining Israel's qualitative military edge. Although Egypt has sought BVR capabilities since the 1970s, its motivation for seeking this capability has likely evolved with the changed threat environment.[91] Originally, Egypt sought this capability to counter Israel; however, as the two countries have moved to strategic cooperation, potential threats from the GERD and the Eastern Mediterranean appear to be the most probable motivations.

Aircraft with longer combat ranges would benefit Egypt because the two closest base options for a potential Egyptian strike on the GERD would be an air base in Aswan, which is around 750 miles from the GERD, or Egypt's Berenice Military Base, which is roughly 850 miles from the GERD (see Figure 3.7).[92] Although Egypt has worked to develop buddy refueling capability, Egypt has traditionally struggled with air-to-air refueling, which constrains the Egyptian Air Force's ability to operate outside its borders.[93] Egypt is only beginning to master probe buddy fueling, not having developed the more efficient boom refueling preferred by many advanced militaries.

Historically, the Egyptian Air Force has primarily relied on F-16s as its most advanced aircraft. However, the United States has consistently prevented Egypt from acquiring air-to-air missiles with a range that exceeds 85 km, which puts Egypt at an "overwhelming disadvantage should it engage in aerial combat with any air force in the region armed with beyond-visual-range (BVR) missiles."[94] Following its move to diversify military acquisitions to lessen dependence on the United States, Egypt ordered 50 MiG-29M/M2 fighter jets from Russia in

Figure 3.7. Geography of a Potential Egyptian Strike on the Grand Ethiopian Renaissance Dam

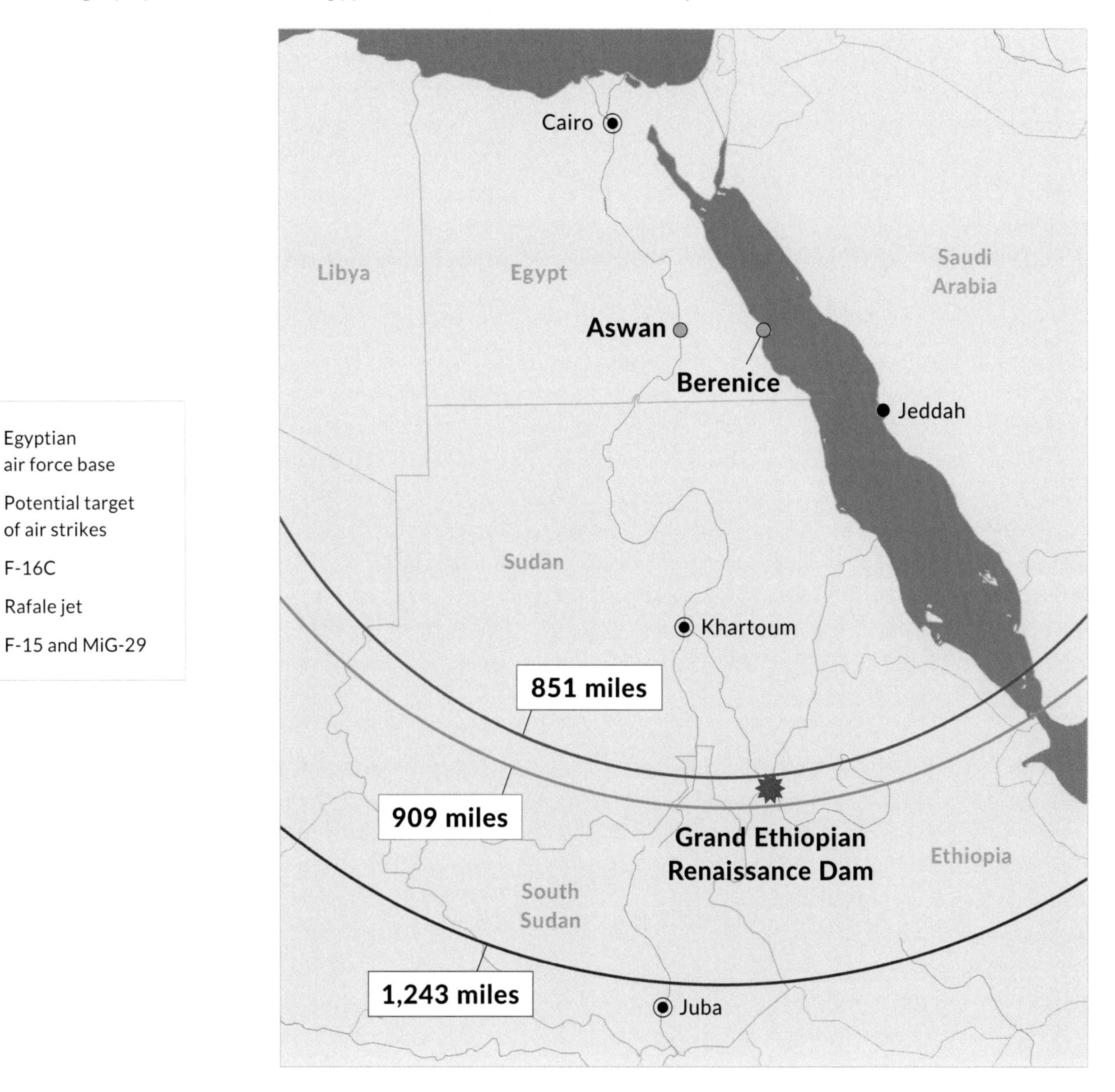

SOURCES: GlobalSecurity.org, "F-16 Fighting Falcon," webpage, undated; vanshilar, "Combat Radius of Western Multirole Fighters," Reddit post, 2016; Peter Suciu, "Russia's MiG-29: Why This Old Fighter Just Won't Go Away," *National Interest*, June 2, 2021; Military Today, "F-15 Advanced Eagle," webpage, undated.

2015 and entered into two deals with France for Rafale fighter jets in 2015 and 2021.[95] It is unclear whether Egypt's Rafales are equipped with MBDA Meteor BVR air-to-air missiles, "but [France] has sold SCALP air-launched cruise missiles and Sagem AASM precision-guided air-to-ground munitions" to Egypt.[96] To remain in compliance with the Missile Technology Control Regime, the range of the Egyptian SCALP missile cannot exceed 300 km.[97] "Egypt also received 300 R-73 and 300 R-77 [BVR] air-to-air missiles for the MiG-29M/M2 fighters."[98] BVR capability may have similarly motivated Egypt's attempt to purchase Russian Su-35s in 2018. However, Egypt abandoned that deal for the 2021 Rafale purchase with France because of potential U.S. sanctions through the Countering America's Adversaries Through Sanctions Act.[99] Most recently, the United States has discussed providing Egypt with F-15s, which Janes has reported could be the Advanced Eagle variant.[100] It is unclear from public reporting whether the United States would provide BVR capability with the F-15s.[101] Supposedly, "industry sources say that Washington has given approval in principle for the sale of the AIM-120 to Cairo.

Raytheon Missiles and Defence, the makers, are currently holding discussions with the Egyptian government."[102]

Egypt's later aircraft acquisitions appear to have longer combat ranges than the country's F-16s; however, several aircraft- and mission-specific factors influence the combat radius. The Rafale is estimated to have a combat range between 485 and 1,025 miles, depending on the mission.[103] For example, the mission radius for a Rafale with three external fuel tanks, "four missiles, [and] four bombs (likely Hammers, 750 lbs)" under optimal cruise conditions would be 790 nautical miles (roughly 909 miles).[104] Both the MiG-29M/M2s and the F-15 Advanced Eagle are reported to have a combat radius of around 1,243 miles.[105] The F-16C's combat radius with two 2,000-lb bombs, two AIM-9 missiles, and one 1,040 U.S. gallon external tank is estimated to be around 851 miles.[106]

Egypt and Sudan have also strengthened their defense ties. For example, the two countries conducted a joint air exercise in March 2021 and an exercise called "Guardians of the Nile" as part of the same series in May 2021.[107] As Ethiopia prepared for the dam's fourth filling stage, despite stalled negotiations over the GERD's water management, Egypt and Sudan conducted joint naval drills at the Port of Sudan.[108] Egypt could potentially seek access to an air base closer to Ethiopia in Sudan if planning an attack.[109] However, Cairo has not initiated any of the processes required to deploy military forces abroad.[110]

For its part, in July 2019, Ethiopia began operating the Spyder Medium Range air defense system near the GERD. Although the system would likely be ineffective in defending against a possible Egyptian attack currently, it could be more effective in the future if integrated into a larger air defense network.[111] The government may also intend for the system to create a "rally around the flag" effect in light of domestic grievances.[112] Later in 2019, Addis Ababa acquired a Pantsir-S1 anti-aircraft defense system that has reportedly been in operation near the GERD.[113] Separately, the commander of the Ethiopian Air Force remarked in June 2021 that "Ethiopian Air Force reinforcements around the Renaissance Dam are stronger than ever."[114] Ethiopia has also acquired Bayraktar TB2 drones that were used in the Tigray conflict in 2021 and Mohajer-6 armed drones from Iran and is potentially negotiating a deal with France regarding other equipment upgrades for its air force.[115] However, apart from the air defense systems, which are more clearly related to a potential GERD conflict, the Tigray conflict could be the primary motivation for these other acquisitions.

Causal Pathway

Figure 3.8 shows the causal pathway from a potential water crisis in Egypt driven by climate change intersecting with the impact of the GERD on Nile River flows.

The first-order consequences are the exacerbation of Egypt's water scarcity and broader changes in Egypt's physical environment. The latter could include a significant rise in salinity in the Nile Delta, with the impact on farming compounded by lower sediment levels being transferred and deposited in this region. Because the Nile Delta is home to an important agri-

Figure 3.8. Potential Causal Pathway to an Egypt–Ethiopia Interstate War

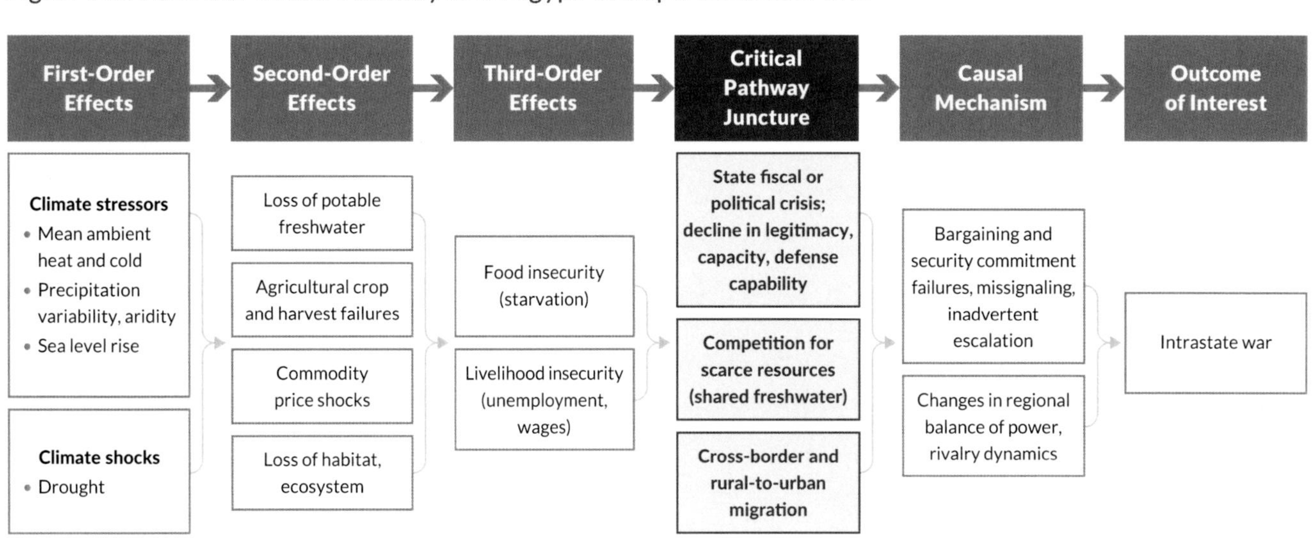

cultural sector, lower productivity could drive up local food prices, contributing to food insecurity. At the same time, those who rely on fresh water in the Nile Delta for their livelihood would also become more food insecure. The overall strain on Egypt could lead to more-generalized economic decline.

Those initial consequences could then be compounded by the migration of people who lose their livelihoods. Cairo's calculus could also change as its environmental challenges threaten to alter the regional balance of power. Cairo might attempt to forestall this shift through the use of military force, either by attempting to compel Ethiopia to limit its filling of the GERD reservoir or by launching a military operation aimed at rallying Egyptians around the flag at a time of heightened internal instability.

Grand Ethiopian Renaissance Dam Conclusion

This case study, which examined the potential for a future Egypt-Ethiopia war over the GERD, illustrates that climate change interacting with additional stress from environmental management could be a basis for interstate war along the western seam of the CENTCOM AOR. To be clear, interstate wars are extremely rare events, and it is unclear under what conditions Egypt may undertake the air strikes on the GERD explored in this analysis. But recent defense acquisitions, Egypt's security cooperation with Sudan, and rhetoric from the countries' respective political and military leaderships suggest that, at a minimum, Egypt and Ethiopia are preparing for such a contingency.

Endnotes

1 Lewis and Lenton, 2015.

2 Omar al-Jaffal and Safaa Khalaf, *'Basra is Burning': The Protests in Basra Governorate, 2018–20*, LSE Middle East Centre, No. 54, October 2021.

3 International Republican Institute, *"Nurid Watan!" "We Want a Homeland!" Basrawi Perspectives on the 2019 Protests in Basra Province*, 2020.

4 Cooley, Ala [يلوك ءالع], "Why Does the Anger of Iraqis Increase During the Summer? [لصف لالخ نييقارعلا بضغ دادزي اذامل ؟فيصلا], Al Jazeera [ةريزجلا ةانق], August 26, 2022.

5 Jennifer Williams, "The Violent Protests in Iraq, Explained," *Vox*, September 8, 2018.

6 Human Rights Watch, *Basra is Thirsty: Iraq's Failure to Manage the Water Crisis*, July 2019.

7 Samya Kullab, "Protesters Block Roads in Iraq After Third Day of Power Cuts," Associated Press, August 8, 2022.

8 Rawabet Center for Strategic Studies [طباورلا زكرم], "Iraq and the Electricity Crisis in the Summer of 2019" ["يف ءابرهكلا ةمزأو قارعلا فيص 2019"], undated; Rawezh Muhamad, "The Electricity Problem in Iraq: From Technical to Political [نم : قارعلا يف ءابرهكلا ةلكشم ةينقت ىلا "ةيسايس"], *Basnews* [لاگنش نسح], June 30, 2021.

9 Alijani Ershad, "No Water or Electricity: Why Southern Iraqis Are at a Breaking Point," *France 24*, July 24, 2018.

10 Human Rights Watch, 2019.

11 Ali Dinar Abdullah, Ilyas Masih, Pieter van der Zaag, Usama F. A. Karim, Ioana Popescu, and Qusay Al Suhail, "Shatt al Arab River System Under Escalating Pressure: A Preliminary Exploration of the Issues and Options for Mitigation," *International Journal of River Basin Management*, Vol. 13, No. 2, April 2015.

12 Human Rights Watch, 2019.

13 "Iran Cuts Electricity Supplies to Iraq over Unpaid Bills," Middle East Monitor, July 7, 2018.

14 Ali Mamouri, "Oil Installations Hit by Protests In Southern Iraq," *Al-Monitor*, July 13, 2018.

15 Human Rights Watch, "Iraq: Security Forces Fire on Protesters," July 2018.

16 Mamouri, 2018; Arwa Ibrahim, "Iraq: Deadly Basra Clashes as Protesters Torch Government Office," Al Jazeera, September 4, 2018.

17 Ahmed Twaij, "Northern Iraq May Be Free, but the South Is Seething," *Foreign Policy*, November 9, 2018.

18 "Protesters in Iraq Set Fires at Iranian Consulate," *New York Times*, September 7, 2018.

19 Hamza Mustafa, "Al-Sistani Supports the Basra Protests and Al-Abadi Tries to Calm Them Down ["يدعم احتجاجات البصرة والعبادي يحاول تهدئتها السيستاني"], *Asharq Al-Awsat* [الشرق الأوسط], July 14, 2018.

20 al-Jaffal and Khalaf, 2021.

21 al-Jaffal and Khalaf, 2021.

22 Michael Knights, "The IRGC May Try to Divert Iraq's Electricity Payments," Washington Institute for Near East Policy, April 5, 2018.

23 "Basra Ignites. . . .What Is Happening in Iraq Now?" ["ةرصبلا لعتشت.. ام يذلا يحدث يف قارعلا ؟نآلا"], *Sasapost* [ةساس تسوب], September 6, 2018.

24 International Republican Institute, 2020.

25 "Basra Ignites. . . . What Is Happening in Iraq Now?" 2018.

26 Ayesha Siddiqi, "Cyclone Bhola: The Disaster That Re-Made South Asia," *Jamhoor*, October 9, 2021; Sravani Biswas and Patrick Daly, "'Cyclone Not Above Politics': East Pakistan, Disaster Politics, and the 1970 Bhola Cyclone," *Modern Asian Studies*, Vol. 55, No. 4, July 2021, p. 1397.

[27] Naomi Hossain, "The 1970 Bhola Cyclone, Nationalist Politics, and the Subsistence Crisis Contract in Bangladesh," *Disasters*, Vol. 42, No. 1, January 2018, p. 191.

[28] Willem van Schendel, *A History of Bangladesh*, Cambridge University Press, 2009, pp. 164, 172.

[29] Institute for the Study of War, "Pakistan and Afghanistan," webpage, undated.

[30] Hasan Zaheer, *The Separation of East Pakistan: The Rise and Realization of Bengali Muslim Nationalism*, Oxford University Press, 1994, p. 16.

[31] Zaheer, 1994, p. 17.

[32] Zaheer, 1994, p. 21.

[33] Zaheer, 1994, p. 38.

[34] Muhammad Iqbal Anjum and Pasquale Michael Sgro, "A Brief History of Pakistan's Economic Development," *Real-World Economics Review*, No. 80, June 2017, p. 173.

[35] Anjum and Sgro, 2017, p. 173.

[36] Zaheer, 1994, p. 52.

[37] Zaheer, 1994, p. 55.

[38] Start Network, *Disaster Summary Sheet: Bangladesh Tropical Storm/ Cyclone,* April 8, 2018.

[39] Start Network, 2018.

[40] Hossain, 2018, p. 191.

[41] Siddiqi, 2021.

[42] Biswas and Daly, 2021, p. 1397.

[43] Siddiqi, 2021.

[44] S. Mahmud Ali, *Understanding Bangladesh*, Columbia University Press, 2010, p. 44.

[45] Biswas and Daly, 2021, p. 1398.

[46] Biswas and Daly, 2021, p. 1400.

[47] Srinath Raghavan, *1971: A Global History of the Creation of Bangladesh*, Harvard University Press, 2013, p. 22.

[48] Raghavan, 2013, pp. 51–52.

[49] Raghavan, 2013, p. 64.

[50] Rodney G. Kyle, *The India–Pakistan War of 1971: A Modern War*, thesis, Marine Corps Command and Staff College, March 14, 1984.

[51] Kyle, 1984.

[52] We are not necessarily forecasting a future war over the GERD. Instead, we are asserting that the main potential combatants are preparing for this contingency because they judge that it may occur.

[53] International Crisis Group, "The Grand Ethiopian Renaissance Dam: A Timeline," June 17, 2020.

[54] John Mukum Mbaku, "The Controversy over the Grand Ethiopian Renaissance Dam," Brookings Institution, August 5, 2020.

[55] "Egypt's el-Sisi Warns 'All Options Open' After Dam Talks Fail," Al Jazeera, April 7, 2021.

[56] Khalil Al-Anani, "The Grand Ethiopian Renaissance Dam: Limited Options for a Resolution," Arab Center Washington DC, September 2022.

[57] Al-Anani, 2022.

[58] "Egypt's Sisi Warns of Potential for Conflict over Ethiopian Dam," Reuters, April 7, 2021.

[59] Beatrice Farhat, "Egypt, Sudan Hold Joint Military Drills as Nile Dispute with Ethiopia Drags On," *Al-Monitor*, April 4, 2023.

[60] Tesfa-Alem Tekle, "Ethiopia Vows to Defend Its Airspace from Any Foreign Attack," *Sudan Tribune*, August 15, 2022.

[61] National Aeronautics and Space Administration, Earth Observatory, "A Grand New Dam on the Nile," webpage, April 19, 2022; World Bank, "Ethiopia's Transformational Approach to Universal Electrification," March 8, 2018.

[62] "Ethiopia's Abiy Inaugurates Electricity Production at Nile Mega-Dam," *France 24*, February 20, 2022.

[63] Ahmed Kamara, Mohamed Ahmed, and Arturo Benavides, "Environmental and Economic Impacts of the Grand Ethiopian Renaissance Dam in Africa," *Water*, Vol. 14, No. 3, January 2022.

[64] Mohammed Basheer, Victor Nechifor, Alvaro Calzadilla, Khalid Siddig, Mikiyas Etichia, Dale Whittington, David Hulme, and Julien J. Harou, "Collaborative Management of the Grand Ethiopian Renaissance Dam Increases Economic Benefits and Resilience," *Nature Communications*, Vol. 12, September 2021.

[65] Hydrologists define *water scarcity* as the existence of less than 1,000 cubic meters of water yearly per capita; Egypt has around 570 cubic meters of water per capita yearly (Ahmad Danburam and Julien Briollais, "GERD, a Path, or Hindrance Toward SDG 6.5 in the Nile River Basin," *Environmental Sciences Proceedings*, Vol. 15, No. 1, May 2022).

[66] Kevin G. Wheeler, Marc Jeuland, Jim W. Hall, Edith Zagona, and Dale Whittington, "Understanding and Managing New Risks on the Nile with the Grand Ethiopian Renaissance Dam," *Nature Communications*, Vol. 11, 2020, p. 2; Basheer et al., 2021.

[67] International Trade Administration, *Egypt—Country Commercial Guide*, U.S. Department of Commerce, August 8, 2022; Dawood Abdelhamid Dawood, "Egypt Non-Conventional Water Resources," video, American University in Cairo, 2022.

[68] Danburam and Briollais, 2022.

[69] Mostafa A. Mohamed, Gamal S. El Afandi, and Mohamed El-Sayed El-Mahdy, "Impact of Climate Change on Rainfall Variability in the Blue Nile Basin," *Alexandria Engineering Journal*, Vol. 61, No. 4, April 2022.

[70] Ethan D. Coffel, Bruce Keith, Corey Lesk, Radley M. Horton, Erica Bower, Jonathan Lee, and Justin S. Mankin, "Future Hot and Dry Years Worsen Nile Basin Water Scarcity Despite Projected Precipitation Increases," *Earth's Future*, Vol. 7, No. 8, August 2019.

[71] Menna Farouk, "Egypt's Farmers Fear Rising Social Tensions over Scarce Water," Reuters, October 31, 2022.

[72] Jean-Daniel Stanley and Pablo Clemente, "Increased Land Subsidence and Sea-Level Rise Are Submerging Egypt's Nile Delta Coastal Margin," *GSA Today*, Vol. 27, No. 5, May 2017, p. 5; Jano Charbel, "Report: Nile Delta's Increasing Salinity and Rising Sea Levels May Make Egypt Uninhabitable by 2100," Mada Masr, March 16, 2017.

[73] Eliora Goodman, "Dual Threats: Water Scarcity and Rising Sea Levels in Egypt," Tahrir Institute for Middle East Policy, August 20, 2021.

[74] Mohammed A. El-Shirbeny and Khaled A. Abutaleb, "Monitoring of Water-Level Fluctuation of Lake Nasser Using Altimetry Satellite Data," *Earth Systems and Environment*, Vol. 2, No. 4, May 2018, p. 367.

[75] Kamara, Ahmed, and Benavides, 2022.

[76] Kevin Wheeler, Marc Jeuland, Kenneth Strzepek, Jim Hall, Edith Zagona, Gamal Abdo, Thinus Basson, Don Blackmore, Paul Block, and Dale Whittington, "Comment on 'Egypt's Water Budget Deficit and Suggested Mitigation Policies for the Grand Ethiopian Renaissance Dam Filling Scenarios,'" *Environmental Research Letters*, Vol. 17, No. 8, August 2022, p. 1.

[77] Danburam and Briollais, 2022.

[78] Wheeler, Jeuland, and Hall, 2020; Gillings School of Global Public Health, "Grand Ethiopian Renaissance Dam Will Make Managing Droughts More Complicated," University of North Carolina, October 16, 2020.

[79] Wheeler, Jeuland, and Hall, 2020, p. 2.

80 Gillings School of Global Public Health, 2020; "Binding Agreement on GERD Required Amid Climate Challenges: Sisi," *Egypt Independent*, November 2022.

81 Wheeler, Jeuland, Strzepek, et al., 2022, p. 11.

82 For an analysis of Egypt's historical approach to air power, see Lon Nordeen and David Nicolle, *Phoenix over the Nile: A History of Egyptian Air Power 1932–1994*, Smithsonian, 1996. Egypt has historical experience with employing air power in Yemen, for example. See Stephanie R. Kelley, *Egypt's Air War in Yemen*, Air Command and Staff College, Air University, March 2010.

83 Ali Dizboni and Karim El-Baz, "Understanding the Egyptian Military's Perspective on the Su-35 Deal," Washington Institute for Near East Policy, July 15, 2021.

84 Michael R. Gordon and David S. Cloud, "U.S. Held Secret Meeting with Israeli, Arab Military Chiefs to Counter Iran Air Threat," *Wall Street Journal*, June 26, 2022.

85 Dizboni and El-Baz, 2021.

86 Dizboni and El-Baz, 2021.

87 Charles W. Dunne, "The Grand Ethiopian Renaissance Dam and Egypt's Military Options," Arab Center Washington DC, July 30, 2020.

88 Dunne, 2020.

89 Dizboni and El-Baz, 2021.

90 Dizboni and El-Baz, 2021.

91 Dizboni and El-Baz, 2021.

92 Dunne, 2020.

93 "Egyptian Fighters Demonstrate Buddy Refueling Capabilities," *Times Aerospace*, undated; Dunne, 2020.

94 Dizboni and El-Baz, 2021.

95 Dizboni and El-Baz, 2021; Darek Liam, "Egyptian Air Force MiG-29 Deliveries Completed, Su-35 Deliveries Begins," Military Africa, July 1, 2021; John Irish, "France to Sell Egypt 30 Fighter Jets in $4.5 Bln Deal—Egyptian Defense Ministry, Report," Reuters, May 3, 2021.

96 Jim Winchester, "Why the Egyptian Air Force Has Such a Varied Fighter Fleet," Key.Aero, July 27, 2022; "Raytheon AMRAAM for Egypt," GBP Aerospace and Defense, March 2022.

97 Jeremy Binnie, "Egyptian Air Force Displays SCALP Cruise Missile," *Janes*, February 3, 2021.

98 Liam, 2021.

99 Liam, 2021.

100 Paul Iddon, "Now that an F-15 Sale Is on the Cards, Will Egypt Give Ukraine Its MiG-29s?" *Forbes*, April 22, 2022a.

101 Paul Iddon, "Egypt Has Spent Big on Diversifying Its Air Force, but to What End?," Middle East Eye, September 9, 2022b.

102 "Raytheon AMRAAM for Egypt," 2022.

103 "A Look at Indian Air Force Rafales' Capabilities as They Are Formally Inducted Today," *Economic Times*, September 10, 2020.

104 vanshilar, 2016.

105 Suciu, 2021; Military Today, undated.

106 GlobalSecurity.org, undated.

107 Mahmoud Gamal, *GERD Crisis: Military Capabilities and Likely Confrontation*, Egyptian Institute for Studies, April 2021; "Photos: Egyptian Forces in Sudan for 'Guardians of the Nile' Military Drill," *Egypt Independent*, May 23, 2021.

108 Farhat, 2023.

109 Shady Ibrahim, "Egypt May Be Looking for a Military Solution to Ethiopia Dam Dispute," Middle East Eye, June 30, 2021; "'Guardians of the Nile' Sudanese and Egyptian Joint Military Exercises—in Pictures," *The National*, June 1, 2021.

110 Ibrahim, 2021; "'Guardians of the Nile' Sudanese and Egyptian Joint Military Exercises—in Pictures," 2021.

111 Karim E. El-Baz, "A Tale of Two Spyders: The Grand Ethiopian Renaissance Dam's Air Defense System or Abiy's Means of Regime Survival," Centre for International and Defence Policy, April 23, 2021.

112 El-Baz, 2021.

113 Gamal, 2021.

114 Mohammed Abdo Hassanein, "Ethiopian Military Buildup around Renaissance Dam," *Asharq Al-Awsat*, June 7, 2021.

115 Ray Mwareya and Ashley Simango, "Turkey's 'Game-Changer' Bayraktar Drones Won't Secure Ethiopia's Shaky Peace," *Newsweek*, March 2, 2022; Ekene Lionel, "Ethiopian Air Force Prepares for War with Egypt," Military Africa, May 31, 2022.

CHAPTER 4

CONCLUSION

THIS REPORT HAS IDENTIFIED causal pathways from climate hazards that lead to different types of conflict, including full-blown intrastate and interstate war. The report has also documented that climate hazards already have—and will plausibly in the future—lead to climate-related conflicts in the CENTCOM AOR.

Our takeaway for readers is that although climate hazards may have direct impact on the incidence of violence, the pathways from climate events to *war* involve multistep processes, in which the initial hazard typically triggers several intervening steps before manifesting in high-intensity conflict. The progression varies but often begins with a climate hazard that results in a form of insecurity (e.g., food, livelihood, physical, or health), which then combines with impacts on state capacity, population flows, and other factors. Then, when filtered through individual and armed group incentives to mobilize around greed or grievance, the pathway culminates in conflict. Readers should also note that there are many variations of the causal pathways from climate hazards to conflict. Our research sought to distill the main variants, an effort that still left us with seven broad families of causal pathways—and many more variants of individual hypotheses—that trace the potential evolution of climate events to conflict.

It is important for the reader to understand that climate hazards interact with other scope conditions (e.g., weak institutional capacity, prior conflict, low economic development) that are more reliable predictors of conflict than exposure to climate hazards. When those scope conditions are reached, climate hazards may operate as a threat multiplier. As will be explored in a forthcoming report that forecasts the frequency of future conflict in the CENTCOM AOR, climate hazards are expected to increase the incidence of conflict but not as significantly as other drivers, including the character of governance, which is intertwined with the level of economic development in the AOR.[1] This is the main limitation of the causal pathways summarized in Chapter 2. Although the pathways explore how climate hazards interact with non-climate variables to produce conflict, the pathways do not focus on the scope conditions (e.g., poor governance, weak economic development, stunted institutional capacity, prior conflict experience) that make the causal pathways more likely.

This report also explored real-world examples of how climate hazards have contributed to conflict onset in the CENTCOM AOR. The overarching findings from this exercise are that climate–related conflict is not just a future phenomenon; it has already occurred in the CENTCOM AOR, contributing to conflict below the threshold of war and to intrastate war. We did not find a compelling case of past climate-driven interstate war in the AOR; however, there are plausible future contingencies, based on our analysis of the defense acquisitions of potential disputants. Our case studies also validated the applicability of the general causal pathways developed in Chapter 2, as those archetypes proved useful in depicting the evolution of the real-world cases we profiled in Chapter 3. The main limitation of the case studies is that our emphasis on analyzing the marginal impact of climate hazards on the outcome can obscure the weight of non-climate factors in explaining the resulting conflict. This is most apparent in the Bhola cyclone case (intrastate war), in which the cyclone was an accelerant, but the main parties already appeared to be on the path to war.

The next step in the research process will be to develop forecasts for the incidence of future conflict in the CENTCOM AOR. This report helps readers understand how climate contributes to conflict onset, whereas the next report will present ranged forecasts for the *frequency* of such conflict onset.[2]

A second forthcoming report will address how three U.S. adversaries—Russia, China, and Iran—may attempt to exploit climate change to advance their interests.[3] A third and final forthcoming report will examine how CENTCOM could use OAIs in the coming decades to address security threats related to climate stress, either by mitigating the risk of climate-related conflict occurring or by responding to such conflicts via military intervention.[4]

Endnotes

1 Toukan et al., 2023.

2 Toukan et al., 2023.

3 Howard J. Shatz, Karen M. Sudkamp, Jeffrey Martini, Mohammad Ahmadi, Derek Grossman, and Kotryna Jukneviciute, *Mischief, Malevolence, or Indifference? How Competitors and Adversaries Could Exploit Climate-Related Conflict in the U.S. Central Command Area of Responsibility*, RAND Corporation, RR-A2338-4, 2023.

4 Karen M. Sudkamp, Elisa Yoshiara, Jeffrey Martini, Mohammad Ahmadi, Matthew Kubasak, Alexander Noyes, Alexandra Stark, Zohan Hasan Tariq, Ryan Haberman, and Erik E. Mueller, *Defense Planning Implications of Climate Change for U.S. Central Command*, RAND Corporation, RR-A2338-5, 2023.

APPENDIX

DIRECT CAUSAL PATHWAYS TO CONFLICT OR VIOLENCE

ALTHOUGH MOST of the hypothetical pathways we identified were indirect, multistep causal links, we also reviewed some 40 peer-reviewed studies that explicitly examined how climate change might directly affect violent interpersonal crime (e.g., murder, assault, rape) or low-level conflict or violence (conceptually defined as fewer than 1,000 battle deaths) via *direct*, one-step causal pathways. We identified three potential one-step pathways, discussed in this appendix, that we summarize in Table A.1. Although direct pathways may be less likely to lead to the large-scale contingencies or crises that CENTCOM plans for, these theoretical pathways are important for identifying future OAIs in at least two ways. First, these direct pathways will likely affect enduring security requirements in the AOR in the future (e.g., military police, installation and diplomatic security). Second, as Homer-Dixon contends, these theories can provide insights into individual-level analyses of the motivations to fight:

> Frustration-aggression theories use individual psychology to explain civil strife, including strikes, riots, coups, revolutions, and guerrilla wars. They suggest that individuals become aggressive when they feel frustrated by something or someone they believe is blocking them from fulfilling a strong desire.[1]

***Hypothesis 8:** Climate change–related increases in ambient temperatures may directly cause increased crime and violence via physiological, neurobiological, and/or psychological mechanisms.*

The first stream in the literature builds on a rich tradition of research that explores potential physiological, neurobiological, and/or psychological links between increases in ambient heat and increases in violent crime or interpersonal violence.[2] As Ranson plainly states, "weather may directly influence people's psychological propensity to commit violent acts of crime."[3] Pre-dating climate-conflict research, proponents of the heat-aggression theory argued, in the words of Anderson, that "clearly, hot temperatures produce increases in aggressive motives and tendencies . . . temperature effects are direct; they operate at the individual level."[4] Anderson later explains the straightforward logic of this argument further:

> Numerous fascinating psychological processes might be involved in the typical effect of high temperatures on aggression and violence. The simplest and most powerful ones all revolve around the "crankiness" notion. Being uncomfortable colors the way people see things. Minor insults may be perceived as major ones, inviting (even demanding) retaliation.[5]

At least as early as 1997, Andersen, Bushman, and Groom began connecting these dots to climate change, positing that "if the heat hypothesis is correct, then global warming should cause increases in violent crime rates."[6] In 2003, Rotton and Cohn are credited with formally articulating the Climate Change–Temperature-Crime Hypothesis, positing in part that violent crime and interpersonal crime will increase because of higher mean ambient temperatures resulting from anthropogenic climate change.[7] Notably, Tiihonen et al. provide the

Table A.1. Summary of Potential Direct Pathways to Intrastate Conflict Below the Threshold of War

Category	Hypothesized Pathway
Direct physiological-biological mechanisms	**Hypothesis 8:** Climate change-related increases in ambient temperatures may directly cause increased crime and violence via physiological, neurobiological, and/or psychological mechanisms.
Direct cognitive mechanisms	**Hypothesis 9:** Climate change-related stressors and shocks may directly increase opportunities for crime via rationale, cognitive cost-benefit analytical mechanisms.
Direct social-behavioral mechanisms	**Hypothesis 10:** Climate change-related stressors and shocks may directly increase opportunities for low-level violence and crime via social-behavioral interaction mechanisms.

first evidence supporting the hypothesis that "causal neurobiological mechanisms" may explain the "association between ambient temperature and aggressive behavior."[8]

Hypothesis 9: *Climate change–related stressors and shocks may directly increase opportunities for crime via rational, cognitive cost-benefit analytical mechanisms.*

The second school of thought—which, as Ranson explains, is rooted in Becker's canonical work of modeling individual criminal decisionmaking on the basis of "rational consideration of the costs and benefits"—posits that "weather is a variable in the production function for crime."[9] According to this logic—as articulated in Ranson (2014) and Jacob, Lefgren, and Moretti (2007)—bad weather (e.g., dark and stormy nights) or extreme weather events may increase opportunities to successfully commit and conceal a violent crime, thereby increasing the frequency of violent crime as climate change raises the incidence of extreme or stormy weather.[10]

Hypothesis 10: *Climate change–related stressors and shocks may directly increase opportunities for low-level violence and crime via social-behavioral interaction mechanisms.*

The third archetypal, direct pathway to low-level interpersonal crime or violence posits that climate change–related weather events—such as unseasonably warm weather spells during periods that are normally cold or vice versa—will lead to increases in crime rates. Building on the "social interaction theory of crime" literature—or the similarly conceived "routine activities theory"—proponents of this theory, such as Rotton and Cohn, argue that climate change–related "improvements" in seasonal or short-term weather patterns may encourage increased social interactions that will lead to temporary spikes in crime rates.[11] By this causal reasoning, assaults might increase during the summer months because people congregate more frequently outside or consume more alcohol at bars or sporting events, thereby increasing the chances of interpersonal conflict and raising the probability of violent encounters. Figure A.1 presents a causal map of the hypothesized direct pathways to low-level violence or conflict.

Figure A.1. Causal Mapping of Hypothesized, Direct Pathways to Low-Level Violence or Conflict

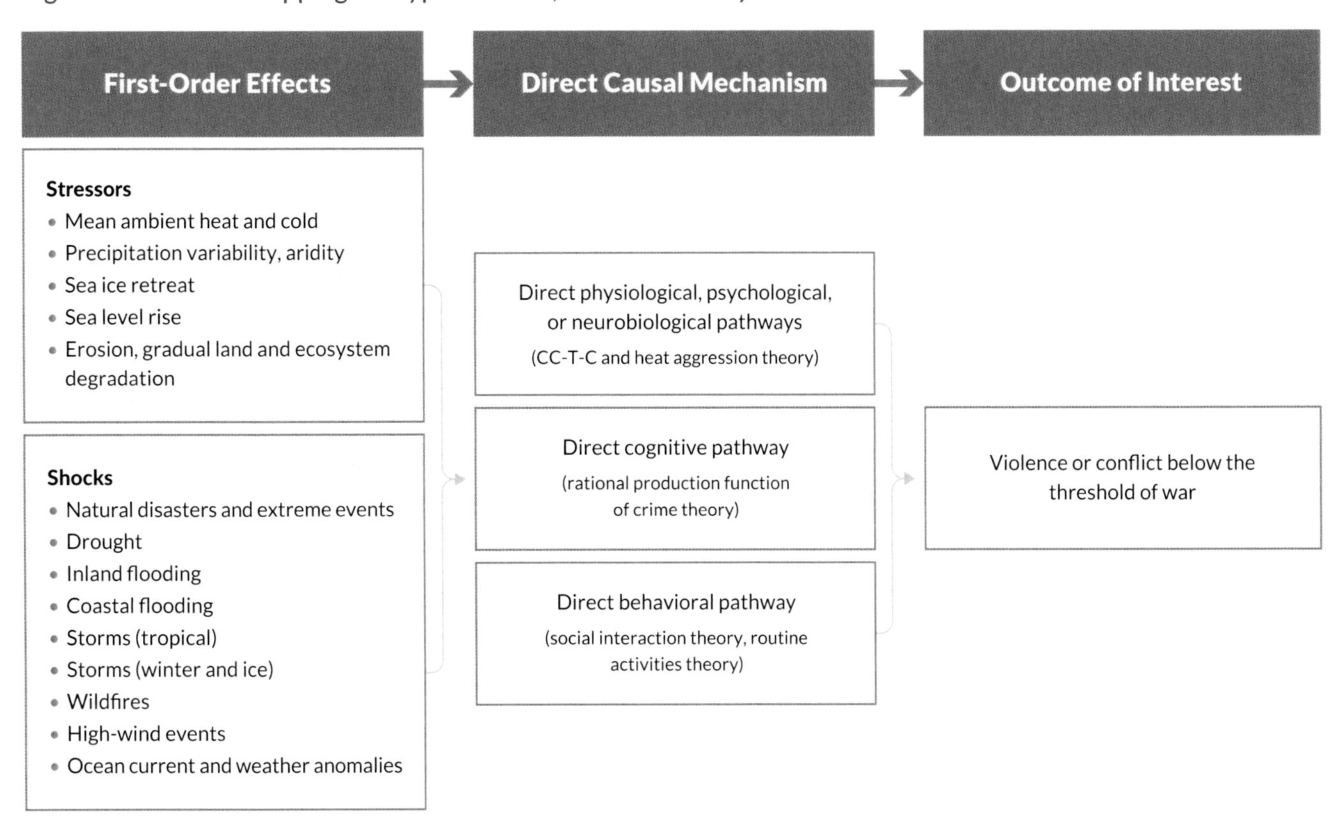

NOTE: CC-T-C = climate change-temperature-crime.

APPENDIX

Endnotes

1 Homer-Dixon, 1999, p. 104.

2 A full retracing of this literature—particularly of 19th century and early-to-mid-20th century works—was beyond the scope of this investigation. However, for a seminal summation, see Craig A. Anderson, "Temperature and Aggression: Ubiquitous Effects of Heat on Occurrence of Human Violence," *Psychological Bulletin*, Vol. 106, No. 1, July 1989, pp. 74, 93. See also Craig A. Anderson, Brad J. Bushman, and Ralph W. Groom, "Hot Years and Serious and Deadly Assault: Empirical Tests of the Heat Hypothesis," *Journal of Personality and Social Psychology*, Vol. 73, No. 6, December 1997, p. 1213; and Craig A. Anderson, "Heat and Violence," *Current Directions in Psychological Science*, Vol. 10, No. 1, February 2001.

3 Matthew Ranson, "Crime, Weather, and Climate Change," Harvard Kennedy School, M-RCBG Associate Working Paper No. 8, May 2012, pp. i, 2, 13–14. See also Matthew Ranson, "Crime, Weather, and Climate Change," *Journal of Environmental Economics and Management*, Vol. 67, No. 3, May 2014.

4 Anderson, 1989, pp. 74, 93–94.

5 Anderson, 2001, p. 36.

6 See Anderson, Bushman, and Groom, 1997, p. 1,215. Allen, Anderson, and Bushman developed a General Affective Aggression Model in an attempt to validate these theorized pathways (Johnie J. Allen, Craig A. Anderson, and Brad J. Bushman, "The General Aggression Model," *Current Opinion in Psychology*, Vol. 19, February 2018). Anderson and Bushman further posit that colder temperatures could also increase aggressiveness and violence through similar, underlying physiological or psychological pathways (Craig A. Anderson and Brad J. Bushman, "Human Aggression," *Annual Review of Psychology*, Vol. 53, No. 1, 2002). See also Craig A. Anderson, Kathryn B. Anderson, Nancy Dorr, Kristina M. DeNeve, and Mindy Flanagan, "Temperature and Aggression," in Mark P. Zanna, ed., *Advances in Experimental Psychology*, Vol. 32, 2000.

7 James Rotton and Ellen G. Cohn, "Global Warming and U.S. Crime Rates: An Application of Routine Activity Theory," *Environment and*

Behavior, Vol. 35, No. 6, November 2003. See also Dennis Mares, "Climate Change and Levels of Violence in Socially Disadvantaged Neighborhood Groups," *Journal of Urban Health*, Vol. 90, No. 4, August 2013; Dennis M. Mares and Kenneth W. Moffett, "Climate Change and Interpersonal Violence: A 'Global' Estimate and Regional Inequities," *Climatic Change*, Vol. 135, No. 2, March 2016, p. 297; Dennis M. Mares and Kenneth W. Moffett, "Climate Change and Crime Revisited: An Exploration of Monthly Temperature Anomalies and UCR Crime Data," *Environment and Behavior*, Vol. 51, No. 5, May 2019, p. 502; David S. Blakeslee and Ram Fishman, "Weather Shocks, Agriculture, and Crime: Evidence from India," *Journal of Human Resources*, Vol. 53, No. 3, Summer 2018; and Robbie M. Parks, James E. Bennett, Helen Tamura-Wicks, Vasilis Kontis, Ralf Toumi, Goodarz Danaei, and Majid Ezzati, "Anomalously Warm Temperatures Are Associated with Increased Injury Deaths," *Nature Medicine*, Vol. 26, No. 1, January 2020.

For an opposing view, see Michael J. Lynch, Paul B. Stretesky, Michael A. Long, and Kimberly L. Barrett, "The Climate Change-Temperature-Crime Hypothesis: Evidence from a Sample of 15 Large U.S. Cities, 2002 to 2015," *International Journal of Offender Therapy and Comparative Criminology*, Vol. 66, No. 4, March 2022.

[8] Jari Tiihonen, Pirjo Halonen, Laura Tiihonen, Hannu Kautiainen, Markus Storvik, and James Callaway, "The Association of Ambient Temperature and Violent Crime," *Scientific Reports*, Vol. 7, July 2017, pp. 1, 5.

[9] Ranson, 2012, p. 3. See also Gary S. Becker, "Crime and Punishment: An Economic Approach," *Journal of Political Economy*, Vol. 76, No. 2, March–April 1968.

[10] Ranson, 2014, p. 276; Brian Jacob, Lars Lefgren, and Enrico Moretti, "The Dynamics of Criminal Behavior: Evidence from Weather Shocks," *Journal of Human Resources*, Vol. 42, No. 3, Summer 2007.

[11] Rotton and Cohn, 2003. According to Ranson (2014), this model of crime is rooted in Glaeser, Sacerdote, and Scheinkman's (1996) social interaction theory of crime. For an application of this theory to the nexus of climate change, see Ranson, 2014, p. 276; and Edward L. Glaeser, Bruce Sacerdote, and José A. Scheinkman, "Crime and Social Interactions," *Quarterly Journal of Economics*, Vol. 111, No. 2, May 1996.

ABBREVIATIONS

AOR	area of responsibility
BVR	beyond-visual-range
CENTCOM	U.S. Central Command
GERD	Grand Ethiopian Renaissance Dam
OAIs	operations, activities, and investments
PPP	Pakistan People's Party
VNSA	violent nonstate actor

REFERENCES

Abdullah, Ali Dinar, Ilyas Masih, Pieter van der Zaag, Usama F. A. Karim, Ioana Popescu, and Qusay Al Suhail, "Shatt al Arab River System Under Escalating Pressure: A Preliminary Exploration of the Issues and Options for Mitigation," *International Journal of River Basin Management*, Vol. 13, No. 2, April 2015.

"A Look at Indian Air Force Rafales' Capabilities as They Are Formally Inducted Today," *Economic Times*, September 10, 2020.

Al-Anani, Khalil, "The Grand Ethiopian Renaissance Dam: Limited Options for a Resolution," Arab Center Washington DC, September 16, 2022.

Ali, S. Mahmud, *Understanding Bangladesh*, Columbia University Press, 2010.

al-Jaffal, Omar, and Safaa Khalaf, *'Basra is Burning': The Protests in Basra Governorate, 2018–20*, LSE Middle East Centre, No. 54, October 2021.

Allen, Johnie J., Craig A. Anderson, and Brad J. Bushman, "The General Aggression Model," *Current Opinion in Psychology*, Vol. 19, February 2018.

Anderson, Craig A., "Temperature and Aggression: Ubiquitous Effects of Heat on Occurrence of Human Violence," *Psychological Bulletin*, Vol. 106, No. 1, July 1989.

Anderson, Craig A., "Heat and Violence," *Current Directions in Psychological Science*, Vol. 10, No. 1, February 2001.

Anderson, Craig A., Kathryn B. Anderson, Nancy Dorr, Kristina M. DeNeve, and Mindy Flanagan, "Temperature and Aggression," in Mark P. Zanna, ed., *Advances in Experimental Social Psychology*, Vol. 32, 2000.

Anderson, Craig A., Brad J. Bushman, and Ralph W. Groom, "Hot Years and Serious and Deadly Assault: Empirical Tests of the Heat Hypothesis," *Journal of Personality and Social Psychology*, Vol. 73, No. 6, December 1997.

Anderson, Craig A., and Brad J. Bushman, "Human Aggression," *Annual Review of Psychology*, Vol. 53, No. 1, 2002.

Angrist, Joshua D., and Adriana D. Kugler, "Rural Windfall or a New Resource Curse? Coca, Income, and Civil Conflict in Colombia," *Review of Economics and Statistics*, Vol. 90, No. 2, May 2008.

Anjum, Muhammad Iqbal, and Pasquale Michael Sgro, "A Brief History of Pakistan's Economic Development," *Real-World Economics Review*, No. 80, June 2017.

Armed Conflict Location and Event Data Project (ACLED), "Data Export Tool," database, last updated June 2023. As of June 27, 2023: https://acleddata.com/data-export-tool/

Ash, Konstantin, and Nick Obradovich, "Climatic Stress, Internal Migration, and Syrian Civil War Onset," *Journal of Conflict Resolution*, Vol. 64, No. 1, January 2020.

Åtland, Kristian, *Security Implications of Climate Change in the Arctic*, Norwegian Defence Research Establishment, No. 2010/01097, May 2010.

Bächler, Günther, "The Anthropogenic Transformation of the Environment: A Source of War?" in Kurt R. Spillmann and Günther Bächler (eds.), *Environmental Crisis: Regional Conflicts and Ways of Cooperation, Environment and Conflicts Project*, No. 14, September 1995.

Bächler, Günther, "Why Environmental Transformation Causes Violence: A Synthesis," *Environmental Change and Security Project Report*, Wilson Center, No. 4, Spring 1998.

Bakaki, Zorzeta, and Roos Haer, "The Impact of Climate Variability on Children: The Recruitment of Boys and Girls by Rebel Groups," *Journal of Peace Research*, Vol. 60, No. 4, July 2023.

Barnett, Jon, "Security and Climate Change," *Global Environmental Change*, Vol. 13, No. 1, April 2003.

Barnett, Jon, and W. Neil Adger, "Climate Change, Human Security, and Violent Conflict," *Political Geography*, Vol 26, No. 6, August 2007.

Bas, Muhammet A., and Elena V. McLean, "Expecting the Unexpected: Disaster Risks and Conflict," *Political Research Quarterly*, Vol. 74, No. 2, June 2021.

Basheer, Mohammed, Victor Nechifor, Alvaro Calzadilla, Khalid Siddig, Mikiyas Etichia, Dale Whittington, David Hulme, and Julien J. Harou, "Collaborative Management of the Grand Ethiopian Renaissance Dam Increases Economic Benefits and Resilience," *Nature Communications*, Vol. 12, September 2021.

"Basra Ignites. . . .What Is Happening in Iraq Now?" ["ةرصبلا لعتشت.. ام يذلا يحدث يف قارعلا نآلا؟"], *Sasapost* [ةساس تسوب], September 6, 2018.

Becker, Gary S., "Crime and Punishment: An Economic Approach," *Journal of Political Economy*, Vol. 76, No. 2, March–April 1968.

Benjaminsen, Tor A., Koffi Alinon, Halvard Buhaug, and Jill Tove Buseth, "Does Climate Change Drive Land-Use Conflicts in the Sahel?" *Journal of Peace Research*, Vol. 49, No. 1, January 2012.

Berkeley Earth, database, undated. As of July 10, 2023: https://berkeleyearth.org/data/

Bernauer, Thomas, and Tobias Böhmelt, "International Conflict and Cooperation over Freshwater Resources," *Nature Sustainability*, Vol. 3, No. 5, May 2020.

Bernauer, Thomas, Tobias Böhmelt, and Vally Koubi, "Environmental Changes and Violent Conflict," *Environmental Research Letters*, Vol. 7, January 2012.

Bernauer, Thomas, and Tobias Siegfried, "Climate Change and International Water Conflict in Central Asia," *Journal of Peace Research*, Vol. 49, No. 1, January 2012.

Biden, Joseph R., *National Security Strategy*, White House, October 2022.

"Binding Agreement on GERD Required Amid Climate Challenges: Sisi," *Egypt Independent*, November 2, 2022.

Binnie, Jeremy, "Egyptian Air Force Displays SCALP Cruise Missile," *Janes*, February 3, 2021.

Biswas, Sravani, and Patrick Daly, "'Cyclone Not Above Politics': East Pakistan, Disaster Politics, and the 1970 Bhola Cyclone," *Modern Asian Studies*, Vol. 55, No. 4, July 2021.

Blakeslee, David S., and Ram Fishman, "Weather Shocks, Agriculture, and Crime: Evidence from India," *Journal of Human Resources,* Vol. 53, No. 3, Summer 2018.

Bogale, Ayalneh, and Benedikt Korf, "To Share or Not to Share? (Non-) Violence, Scarcity, and Resource Access in Somali Region, Ethiopia," *Journal of Development Studies*, Vol. 43, No. 4, May 2007.

Borgerson, Scott G., "Arctic Meltdown: The Economic and Security Implications of Global Warming," *Foreign Affairs*, Vol. 87, No. 2, March–April 2008.

Brochmann, Marit, and Nils Petter Gleditsch, "Shared Rivers and Conflict—A Reconsideration," *Political Geography*, Vol. 31, No. 8, November 2012.

Brzoska, Michael, and Christiane Fröhlich, "Climate Change, Migration, and Violent Conflict: Vulnerabilities, Pathways, and Adaptation Strategies," *Migration and Development*, Vol. 5, No. 2, 2016.

Buhaug, Halvard, Tor A. Benjaminsen, Espen Sjaastad, and Ole Magnus Theisen, "Climate Variability, Food Production Shocks, and Violent Conflict in Sub-Saharan Africa," *Environmental Research Letters*, Vol. 10, No. 12, December 2015.

Bunn, Matthew, "Nuclear Disarmament, Nuclear Energy, and Climate Change: Exploring the Linkages," in Bård Nikolas Vik Steen and Olav Njølstad, eds., *Nuclear Disarmament: A Critical Assessment*, Routledge, 2019.

Burrows, Kate, and Patrick L. Kinney, "Exploring the Climate Change, Migration and Conflict Nexus," *International Journal of Environmental Research and Public Health*, Vol. 13, No. 4, April 2016.

Butler, Christopher K., and Scott Gates, "African Range Wars: Climate, Conflict, and Property Rights," *Journal of Peace Research*, Vol. 49, No. 1, January 2012.

Cattaneo, Cristina, and Timothy Foreman, *Climate Change, International Migration, and Interstate Conflict*, Centre for Research and Analysis of Migration, CDP 09/21, March 31, 2021.

Charbel, Jano, "Report: Nile Delta's Increasing Salinity and Rising Sea Levels May Make Egypt Uninhabitable by 2100," Mada Masr, March 16, 2017.

Church, Clare, and Alec Crawford, "Minerals and the Metals for the Energy Transition: Exploring the Conflict Implications for Mineral-Rich, Fragile States," in Manfred Hafner and Simone Tagliapietra, eds., *The Geopolitics of the Global Energy Transition*, Lecture Notes in Energy, Vol. 73, Springer, 2020.

Coffel, Ethan D., Bruce Keith, Corey Lesk, Radley M. Horton, Erica Bower, Jonathan Lee, and Justin S. Mankin, "Future Hot and Dry Years Worsen Nile Basin Water Scarcity Despite Projected Precipitation Increases," *Earth's Future*, Vol. 7, No. 8, August 2019.

Cooley, Ala [علاء كولي], "Why Does the Anger of Iraqis Increase During the Summer?" ["لماذا يزداد غضب العراقيين خلال فصل الصيف؟"], Al Jazeera [قناة الجزيرة], August 26, 2022.

Danburam, Ahmad, and Julien Briollais, "GERD, a Path, or Hindrance Toward SDG 6.5 in the Nile River Basin?" *Environmental Sciences Proceedings*, Vol. 15, No. 1, May 2022.

Dawood, Dawood Abdelhamid, "Egypt Non-Conventional Water Resources," video, American University in Cairo, 2021. As of March 9, 2023:
https://fount.aucegypt.edu/audiovisual_faculty_work/6/

De Châtel, Francesca, "The Role of Drought and Climate Change in the Syrian Uprising: Untangling the Triggers of the Revolution," *Middle Eastern Studies*, Vol. 50, No. 4, May 2014.

Devlin, Colleen, and Cullen S. Hendrix, "Trends and Triggers Redux: Climate Change, Rainfall, and Interstate Conflict," *Political Geography*, Vol. 43, November 2014.

Dinar, Shlomi, David Katz, Lucia De Stefano, and Brian Blankespoor, "Climate Change, Conflict, and Cooperation: Global Analysis of the Effectiveness of International River Treaties in Addressing Water Variability," *Political Geography*, Vol. 45, March 2015.

Dizboni, Ali, and Karim El-Baz, "Understanding the Egyptian Military's Perspective on the Su-35 Deal," Washington Institute for Near East Policy, July 15, 2021.

Dube, Oeindrila, and Juan F. Vargas, "Commodity Price Shocks and Civil Conflict: Evidence from Colombia," *Review of Economic Studies*, Vol. 80, No. 4, October 2013.

Dunne, Charles W., "The Grand Ethiopian Renaissance Dam and Egypt's Military Options," Arab Center Washington DC, July 30, 2020.

Dupont, Alan, and Graeme Pearman, *Heating Up the Planet: Climate Change and Security*, Lowy Institute, Paper No. 12, June 2006.

Eastin, Joshua, "Fuel to the Fire: Natural Disasters and the Duration of Civil Conflict," *International Interactions*, Vol. 42, No. 2, February 2016.

"Egypt's el-Sisi Warns 'All Options Open' After Dam Talks Fail," Al Jazeera, April 7, 2021.

"Egypt's Sisi Warns of Potential for Conflict over Ethiopian Dam," Reuters, April 7, 2021.

"Egyptian Fighters Demonstrate Buddy Refueling Capabilities," *Times Aerospace*, undated.

Eisgruber, Lasse, "The Resource Curse: Analysis of the Applicability to the Large-Scale Export of Electricity from Renewable Resources," *Energy Policy*, Vol. 57, June 2013.

El-Baz, Karim E., "A Tale of Two Spyders: The Grand Ethiopian Renaissance Dam's Air Defense System or Abiy's Means of Regime Survival," Centre for International and Defence Policy, April 23, 2021.

El-Shirbeny, Mohammed A., and Khaled A. Abutaleb, "Monitoring of Water-Level Fluctuation of Lake Nasser Using Altimetry Satellite Data," *Earth Systems and Environment*, Vol. 2, No. 4, May 2018.

Ember, Carol R., Teferi Abate Adem, Ian Skoggard, and Eric C. Jones, "Livestock Raiding and Rainfall Variability in Northwestern Kenya," *Civil Wars*, Vol. 14, No. 2, June 2012.

Ember, Carol R., Ian Skoggard, Teferi Abate Adem, and A. J. Faas, "Rain and Raids Revisited: Disaggregating Ethnic Group Livestock Raiding in the Ethiopian-Kenyan Border Region," *Civil Wars*, Vol. 16, No. 3, July 2014.

Ershad, Alijani, "No Water or Electricity: Why Southern Iraqis Are at a Breaking Point," *France 24*, July 24, 2018.

"Ethiopia's Abiy Inaugurates Electricity Production at Nile Mega-Dam," *France 24*, February 20, 2022.

REFERENCES

Eyl-Mazzega, Marc-Antoine, and Carole Mathieu, "The European Union and the Energy Transition," in Manfred Hafner and Simone Tagliapietra, eds., *The Geopolitics of the Global Energy Transition*, Lecture Notes in Energy, Vol. 73, Springer, 2020.

Farhat, Beatrice, "Egypt, Sudan Hold Joint Military Drills as Nile Dispute with Ethiopia Drags On," *Al-Monitor*, April 4, 2023.

Farouk, Menna, "Egypt's Farmers Fear Rising Social Tensions over Scarce Water," Reuters, October 31, 2022.

Fjelde, Hanne, and Nina von Uexkull, "Climate Triggers: Rainfall Anomalies, Vulnerability and Communal Conflict in Sub-Saharan Africa," *Political Geography*, Vol. 31, No. 7, September 2012.

Fröhlich, Christiane, and Giovanna Gioli, "Gender, Conflict, and Global Environmental Change," *Peace Review*, Vol. 27, No. 2, 2015.

Gamal, Mahmoud, *GERD Crisis: Military Capabilities and Likely Confrontation*, Egyptian Institute for Studies, April 2021.

Gemenne, François, Jon Barnett, W. Neil Adger, and Geoffrey D. Dabelko, "Climate and Security: Evidence, Emerging Risks, and a New Agenda," *Climatic Change*, Vol. 123, No. 1, March 2014.

Gillings School of Global Public Health, "Grand Ethiopian Renaissance Dam Will Make Managing Droughts More Complicated," University of North Carolina, October 16, 2020.

Glaeser, Edward L., Bruce Sacerdote, and José A. Scheinkman, "Crime and Social Interactions," *Quarterly Journal of Economics*, Vol. 111, No. 2, May 1996.

Gleditsch, Nils Petter, "This Time Is Different! Or Is It? NeoMalthusians and Environmental Optimists in the Age of Climate Change," *Journal of Peace Research*, Vol. 58, No. 1, January 2021.

Gleditsch, Nils Petter, Kathryn Furlong, Håvard Hegre, Bethany Lacina, and Taylor Owen, "Conflicts over Shared Rivers: Resource Scarcity or Fuzzy Boundaries?" *Political Geography*, Vol. 25, No. 4, May 2006.

Gleditsch, Nils Petter, and Henrik Urdal, "Ecoviolence? Links Between Population Growth, Environmental Scarcity and Violent Conflict in Thomas Homer-Dixon's Work," *Journal of International Affairs*, Vol. 56, No. 1, Fall 2002.

Gleick, Peter H., "Water and Conflict: Fresh Water Resources and International Security," *International Security*, Vol. 18, No. 1, Summer 1993.

Gleick, Peter H., "Water, Drought, Climate Change, and Conflict in Syria," *Weather, Climate, and Society*, Vol. 6, No. 3, July 2014.

GlobalSecurity.org, "F-16 Fighting Falcon," webpage, undated. As of March 9, 2023: https://www.globalsecurity.org/military/systems/aircraft/f-16-specs.htm

Goodman, Eliora, "Dual Threats: Water Scarcity and Rising Sea Levels in Egypt," Tahrir Institute for Middle East Policy, August 20, 2021.

Gordon, Michael R., and David S. Cloud, "U.S. Held Secret Meeting with Israeli, Arab Military Chiefs to Counter Iran Air Threat," *Wall Street Journal*, June 26, 2022.

"'Guardians of the Nile' Sudanese and Egyptian Joint Military Exercises—in Pictures," *The National*, June 1, 2021.

Habib, Komal, Lorie Hamelin, and Henrik Wenzel, "A Dynamic Perspective of the Geopolitical Supply Risk of Metals," *Journal of Cleaner Production*, Vol. 133, October 2016.

Hassanein, Mohammed Abdo, "Ethiopian Military Buildup Around Renaissance Dam," *Asharq Al-Awsat*, June 7, 2021.

Hendrix, Cullen S., "Oil Prices and Interstate Conflict," *Conflict Management and Peace Science*, Vol. 34, No. 6, November 2017.

Hendrix, Cullen S., and Idean Salehyan, "Climate Change, Rainfall, and Social Conflict in Africa," *Journal of Peace Research*, Vol. 49, No. 1, January 2012.

Hensel, Paul R., Sara McLaughlin Mitchell, and Thomas E. Sowers II, "Conflict Management of Riparian Disputes," *Political Geography*, Vol. 25, No. 4, May 2006.

Heslin, Alison, "Riots and Resources: How Food Access Affects Collective Violence," *Journal of Peace Research*, Vol. 58, No. 2, April 2020.

Homer-Dixon, Thomas, "On the Threshold: Environmental Changes as Causes of Acute Conflict," *International Security*, Vol. 16, No. 2, Fall 1991.

Homer-Dixon, Thomas, "Environmental Scarcities and Violent Conflict: Evidence from Cases," *International Security*, Vol. 19, No. 1, Summer 1994.

Homer-Dixon, Thomas, "The Project on Environment, Population and Security: Key Findings of Research," *Environmental Change and Security Project Report 2*, Woodrow Wilson Center, 1996.

Homer-Dixon, Thomas, *Environment, Scarcity, and Violence*, Princeton University Press, 1999.

Homer-Dixon, Thomas, "Terror in the Weather Forecast," *New York Times*, April 24, 2007.

Homer-Dixon, Thomas, and Jessica Blitt, "Introduction: A Theoretical Overview," in Thomas Homer-Dixon and Jessica Blitt, eds., *Ecoviolence: Links Among Environment, Population, and Security*, Rowman and Littlefield, 1998.

Hossain, Naomi, "The 1970 Bhola Cyclone, Nationalist Politics, and the Subsistence Crisis Contract in Bangladesh," *Disasters*, Vol. 42, No. 1, January 2018.

Human Rights Watch, "Iraq: Security Forces Fire on Protesters," July 24, 2018.

Human Rights Watch, *Basra is Thirsty: Iraq's Failure to Manage the Water Crisis*, July 2019.

Ibrahim, Arwa, "Iraq: Deadly Basra Clashes as Protesters Torch Government Office," Al Jazeera, September 4, 2018.

Ibrahim, Shady, "Egypt May Be Looking for a Military Solution to Ethiopia Dam Dispute," Middle East Eye, June 30, 2021.

Iddon, Paul, "Now That an F-15 Sale Is on the Cards, Will Egypt Give Ukraine Its MiG-29s?" *Forbes*, April 22, 2022a.

Iddon, Paul, "Egypt Has Spent Big on Diversifying Its Air Force, but to What End?" Middle East Eye, September 9, 2022b.

Ide, Tobias, "Research Methods for Exploring the Links Between Climate Change and Conflict," *WIREs Climate Change*, Vol. 8, No. 3, May–June 2017.

Ide, Tobias, Michael Brzoska, Jonathan F. Donges, and Carl-Friedrich Schleussner, "Multi-Method Evidence for When and How Climate-Related Disasters Contribute to Armed Conflict Risk," *Global Environmental Change*, Vol. 62, May 2020.

Institute for the Study of War, "Pakistan and Afghanistan," webpage, undated. As of March 9, 2023: https://www.understandingwar.org/pakistan-and-afghanistan

International Crisis Group, "The Grand Ethiopian Renaissance Dam: A Timeline," June 17, 2020.

International Energy Agency, "Iraq's Electricity Supply and Demand, 2018–2030," webpage, last updated April 25, 2019. As of June 22, 2023:

https://www.iea.org/data-and-statistics/charts/iraqs-electricity-supply-and-demand-2018-2030

International Republican Institute, *"Nurid Watan!" "We Want a Homeland!" Basrawi Perspectives on the 2019 Protests in Basra Province*, 2020.

International Trade Administration, *Egypt—Country Commercial Guide*, U.S. Department of Commerce, August 8, 2022.

"Iran Cuts Electricity Supplies to Iraq over Unpaid Bills," Middle East Monitor, July 7, 2018.

Irish, John, "France to Sell Egypt 30 Fighter Jets in $4.5 Bln Deal—Egyptian Defense Ministry, Report," Reuters, May 3, 2021.

Iyigun, Murat, Nathan Nunn, and Nancy Qian, *Winter is Coming: The Long-Run Effects of Climate Change on Conflict, 1400–1900*, IZA Institute of Labor Economics, No. 10475, January 2017.

Jacob, Brian, Lars Lefgren, and Enrico Moretti, "The Dynamics of Criminal Behavior: Evidence from Weather Shocks," *Journal of Human Resources*, Vol. 42, No. 3, Summer 2007.

Jones, Benjamin T., Eleonora Mattiacci, and Bear F. Braumoeller, "Food Scarcity and State Vulnerability: Unpacking the Link Between Climate Variability and Violent Unrest," *Journal of Peace Research*, Vol. 54, No. 3, May 2017.

Kahl, Colin H., *States, Scarcity, and Civil Strife in the Developing World*, Princeton University Press, 2006.

Kalbhenn, Anna, "Liberal Peace and Shared Resources—A Fair-Weather Phenomenon?" *Journal of Peace Research*, Vol. 48, No. 6, November 2011.

Kallis, Giorgos, and Christos Zografos, "Hydro-Climatic Change, Conflict and Security," *Climatic Change*, Vol. 123, No. 1, March 2014.

Kamara, Ahmed, Mohamed Ahmed, and Arturo Benavides, "Environmental and Economic Impacts of the Grand Ethiopian Renaissance Dam in Africa," *Water*, Vol. 14, No. 3, January 2022.

Karnieli, Arnon, Alexandra Shtein, Natalya Panov, Noam Weisbrod, and Alon Tal, "Was Drought Really the Trigger Behind the Syrian Civil War in 2011?" *Water*, Vol. 11, No. 8, July 2019.

Kelley, Colin P., Shahrzad Mohtadi, Mark A. Cane, Richard Seager, and Yochanan Kushnir, "Climate Change in the Fertile Crescent and Implications of the Recent Syrian Drought," *Proceedings of the National Academy of Sciences*, Vol. 112, No. 11, March 17, 2015.

Kelley, Stephanie R., *Egypt's Air War in Yemen*, Air Command and Staff College, Air University, March 2010.

Kingdon, Ashton, and Briony Gray, "The Class Conflict Rises When You Turn up the Heat: An Interdisciplinary Examination of the Relationship Between Climate Change and Left-Wing Terrorist Recruitment," *Terrorism and Political Violence*, Vol. 34, No. 5, May 2022.

Knights, Michael, "The IRGC May Try to Divert Iraq's Electricity Payments," Washington Institute for Near East Policy, April 5, 2018.

Koubi, Vally, "Climate Change and Conflict," *Annual Review of Political Science*, Vol. 22, May 2019.

Koubi, Vally, Tobias Böhmelt, Gabriele Spilker, and Lena Schaffer, "The Determinants of Environmental Migrants' Conflict Perception," *International Organization*, Vol. 72, No. 4, Fall 2018.

Koubi, Vally, Gabriele Spilker, Tobias Böhmelt, and Thomas Bernauer, "Do Natural Resources Matter for Interstate and Intrastate Armed Conflict?" *Journal of Peace Research*, Vol. 51, No. 2, March 2014.

Kullab, Samya, "Protesters Block Roads in Iraq After Third Day of Power Cuts," Associated Press, August 8, 2022.

Kyle, Rodney G., *The India–Pakistan War of 1971: A Modern War*, thesis, Marine Corps Command and Staff College, March 14, 1984.

Lee, Bomi K., Sara McLaughlin Mitchell, Cody J. Schmidt, and Yufan Yang, "Disasters and the Dynamics of Interstate Rivalry," *Journal of Peace Research*, Vol. 59, No. 1, March 2022.

Lee, James R., and Kisei R. Tanaka, "Climate Change, Conflict, and Moving Borders," *International Journal of Climate Change Impacts and Responses*, Vol. 8, No. 3, January 2016.

Leonard, Alycia, Aniq Ahsan, Flora Charbonnier, and Stephanie Hirmer, "The Resource Curse in Renewable Energy: A Framework for Risk Assessment," *Energy Strategy Reviews*, Vol. 41, May 2022.

Lewis, Kristy H., and Timothy M. Lenton, "Knowledge Problems in Climate Change and Security Research," *WIREs Climate Change*, Vol. 6, No. 4, July–August 2015.

Liam, Darek, "Egyptian Air Force MiG-29 Deliveries Completed, Su-35 Deliveries Begins," Military Africa, July 1, 2021.

Linke, Andrew M., and Brett Ruether, "Weather, Wheat, and War: Security Implications of Climate Variability for Conflict in Syria," *Journal of Peace Research*, Vol. 58, No. 1, January 2021.

Lionel, Ekene, "Ethiopian Air Force Prepares for War with Egypt," Military Africa, May 31, 2022.

Lynch, Michael J., Paul B. Stretesky, Michael A. Long, and Kimberly L. Barrett, "The Climate Change-Temperature-Crime Hypothesis: Evidence from a Sample of 15 Large U.S. Cities, 2002 to 2015," *International Journal of Offender Therapy and Comparative Criminology*, Vol. 66, No. 4, March 2022.

Lyons, Scott W., "Preventing a Renewable Resource Curse," *Sustainable Development Law and Policy*, Vol. 15, No. 2, Spring 2015.

Malji, Andrea, Laurabell Obana, and Cidney Hopkins, "When Home Disappears: South Asia and the Growing Risk of Climate Conflict," *Terrorism and Political Violence*, Vol. 34, No. 5, May 2022.

Malthus, Thomas, *An Essay on the Principle of Population*, London, 1798.

Mamouri, Ali, "Oil Installations Hit by Protests in Southern Iraq," *Al-Monitor*, July 13, 2018.

Månberger, André, and Bengt Johansson, "The Geopolitics of Metals and Metalloids Used for the Renewable Energy Transition," *Energy Strategy Reviews*, Vol. 26, November 2019.

Månsson, André, "A Resource Curse for Renewables? Conflict and Cooperation in the Renewable Energy Sector," *Energy Research and Social Science*, Vol. 10, November 2015.

Mares, Dennis, "Climate Change and Levels of Violence in Socially Disadvantaged Neighborhood Groups," *Journal of Urban Health*, Vol. 90, No. 4, August 2013.

Mares, Dennis M., and Kenneth W. Moffett, "Climate Change and Interpersonal Violence: A 'Global' Estimate and Regional Inequities," *Climatic Change*, Vol. 135, No. 2, March 2016.

Mares, Dennis M., and Kenneth W. Moffett, "Climate Change and Crime Revisited: An Exploration of Monthly Temperature Anomalies and UCR Crime Data," *Environment and Behavior*, Vol. 51, No. 5, May 2019.

Markowitz, Jonathan N., *Perils of Plenty: Arctic Resource Competition and the Return of the Great Game*, Oxford University Press, 2020.

Mbaku, John Mukum, "The Controversy over the Grand Ethiopian Renaissance Dam," Brookings Institution, August 5, 2020.

McDonald, Matt, "Discourses of Climate Security," *Political Geography*, Vol. 33, March 2013.

Meier, Patrick, Doug Bond, and Joe Bond, "Environmental Influences on Pastoral Conflict in the Horn of Africa," *Political Geography*, Vol. 26, No. 6, August 2007.

Meiners, Roger E., and Andrew P. Morris, *Addressing Green Energy's "Resource Curse,"* Texas A&M University School of Law, No. 22–31, February 16, 2022.

Military Advisory Board, *National Security and the Threat of Climate Change,* CNA Corporation, 2007.

Military Today, "F-15 Advanced Eagle," webpage, undated. As of September 27, 2023: https://www.militarytoday.com/aircraft/f15_advanced_eagle.htm

Miro, Michelle E., Flannery Dolan, Karen M. Sudkamp, Jeffrey Martini, Karishma V. Patel, and Carlos Calvo Hernandez, *A Hotter and Drier Future Ahead An Assessment of Climate Change in U.S. Central Command*, RAND Corporation, RR-A2338-1, 2023.

Mohamed, Mostafa A., Gamal S. El Afandi, and Mohamed El-Sayed El-Mahdy, "Impact of Climate Change on Rainfall Variability in the Blue Nile Basin," *Alexandria Engineering Journal*, Vol. 61, No. 4, April 2022.

Muhamad, Rawezh, "The Electricity Problem in Iraq: From Technical to Political" ["ةيسايس" ىلا ةينقت نم : قارعلا يف ءابرهكلا ةلكشم"], *Basnews* [لاگنش نسح], June 30, 2021.

Mustafa, Hamza, "Al-Sistani Supports the Basra Protests and Al-Abadi Tries to Calm Them Down ["يدعم احتجاجات البصرة والعبادي يحاول تهدئتها السيستاني"], *Asharq Al-Awsat* [الشرق الأوسط], July 14, 2018.

Mwareya, Ray, and Ashley Simango, "Turkey's 'Game-Changer' Bayraktar Drones Won't Secure Ethiopia's Shaky Peace," *Newsweek*, March 2, 2022.

Nguyen, Huong Thu, "Gendered Vulnerabilities in Times of Natural Disasters: Male-to-Female Violence in the Philippines in the Aftermath of Super Typhoon Haiyan," *Violence Against Women*, Vol. 25, No. 4, August 2018.

National Aeronautics and Space Administration, Earth Observatory, "A Grand New Dam on the Nile," webpage, April 19, 2022. As of March 9, 2023: https://earthobservatory.nasa.gov/images/149691/a-grand-new-dam-on-the-nile

Nordeen, Lon, and David Nicolle, *Phoenix over the Nile: A History of Egyptian Air Power 1932–1994*, Smithsonian, 1996.

O'Sullivan, Megan, Indra Overland, and David Sandalow, *The Geopolitics of Renewable Energy*, Columbia University Center on Global Energy Policy and Harvard Kennedy School Belfer Center for Science and International Affairs, June 2017.

Papaioannou, Kostadis J., "Climate Shocks and Conflict: Evidence from Colonial Nigeria," *Political Geography*, Vol. 50, January 2016.

Parks, Robbie M., James E. Bennett, Helen Tamura-Wicks, Vasilis Kontis, Ralf Toumi, Goodarz Danaei, and Majid Ezzati, "Anomalously Warm Temperatures Are Associated with Increased Injury Deaths," *Nature Medicine*, Vol. 26, No. 1, January 2020.

Parthemore, Christine, Francesco Femia, and Caitlin Werrell, "The Global Responsibility to Prepare for Intersecting Climate and Nuclear Risks," *Bulletin of the Atomic Scientists*, Vol. 74, No. 6, 2018.

Pelling, Mark, and Kathleen Dill, "Disaster Politics: Tipping Points for Change in the Adaptation of Sociopolitical Regimes," *Progress in Human Geography*, Vol. 34, No. 1, February 2010.

"Photos: Egyptian Forces in Sudan for 'Guardians of the Nile' Military Drill," *Egypt Independent*, May 23, 2021.

Podesta, John, and Peter Ogden, "The Security Implications of Climate Change," *Washington Quarterly,* Vol. 31, No. 1, January 2008.

Pörtner, Hans-Otto, Debra C. Roberts, Melinda M. B. Tignor, Elvira Poloczanska, Katja Mintenbeck, Andrés Alegría, Marlies Craig, Stefanie Langsdorf, Sina Löschke, Vincent Möller, Andrew Okem, and Bardhyl Rama, eds., *Climate Change 2022: Impacts, Adaptation, and Vulnerability. Working Group II Contribution to the Sixth Assessment Report of the Intergovernmental Panel on Climate Change*, Intergovernmental Panel on Climate Change, Cambridge University, 2022.

"Protesters in Iraq Set Fires at Iranian Consulate," *New York Times*, September 7, 2018.

Raghavan, Srinath, *1971: A Global History of the Creation of Bangladesh*, Harvard University Press, 2013.

Raleigh, Clionadh, and Dominic Kniveton, "Come Rain or Shine: An Analysis of Conflict and Climate Variability in East Africa," *Journal of Peace Research*, Vol. 49, No. 1, January 2012.

Ranson, Matthew, "Crime, Weather, and Climate Change," Harvard Kennedy School, M-RCBG Associate Working Paper No. 8, May 2012.

Ranson, Matthew, "Crime, Weather, and Climate Change," *Journal of Environmental Economics and Management*, Vol. 67, No. 3, May 2014.

Ravnborg, Helle Munk, Rocio Bustamante, Abdoulaye Cissé, Signe M. Cold-Ravnkilde, Vladimir Cossio, Moussa Djiré, Mikkel Funder, Ligia I. Gómez, Phuong Le, Carol Mweemba, Imasiku Nyambe, Tania Paz, Huong Pham, Roberto Rivas, Thomas Skielboe, and Nguyen T. B. Yen, "Challenges of Local Water Governance: The Extent, Nature, and Intensity of Local Water-Related Conflict and Cooperation," *Water Policy*, Vol. 14, No. 2, April 2012.

Rawabet Center for Strategic Studies [طباورلا زكرم], "Iraq and the Electricity Crisis in the Summer of 2019" ["يف ءابرهكلا ةمزأو قارعلا فيص 2019"]," undated.

"Raytheon AMRAAM for Egypt," GBP Aerospace and Defense, March 21, 2022.

Reuveny, Rafael, "Climate Change-Induced Migration and Violent Conflict," *Political Geography*, Vol. 26, No. 6, August 2007.

Reuveny, Rafael, "Ecomigration and Violent Conflict: Case Studies and Public Policy Implications," *Human Ecology*, Vol. 36, No. 1, February 2008.

Rotton, James, and Ellen G. Cohn, "Global Warming and U.S. Crime Rates: An Application of Routine Activity Theory," *Environment and Behavior*, Vol. 35, No. 6, November 2003.

Sakaguchi, Kendra, Anil Varughese, and Graeme Auld, "Climate Wars? A Systematic Review of Empirical Analyses on the Links Between Climate Change and Violent Conflict," *International Studies Review*, Vol. 19, No. 4, December 2017.

Salehyan, Idean, "From Climate Change to Conflict? No Consensus Yet," *Journal of Peace Research*, Vol. 45, No. 3, May 2008.

Sanz-Barbero, Belén, Cristina Linares, Carmen Vives-Cases, José Luis González, Juan José López-Ossorio, and Julio Díaz, "Heat Wave and the Risk of Intimate Partner Violence," *Science of the Total Environment*, Vol. 644, December 2018.

Sarkees, Meredith Reid, *The COW Typology of War: Defining and Categorizing Wars (Version 4 of the Data)*, Correlates of War, undated.

Scheffran, Jürgen, and Antonella Battaglini, "Climate and Conflicts: The Security Risks of Global Warming," *Regional Environmental Change*, Vol. 11, Supp. 1, March 2011.

Scheffran, Jürgen, P. Michael Link, and Janpeter Schilling, "Theories and Models of the Climate-Security Interaction: Framework and Application to a Climate Hot Spot in North Africa," in Jürgen Scheffran, Michael Brzoska, Hans Günter Brauch, Peter Michael Link, and Janpeter Schilling, eds., *Climate Change, Human Security and Violent Conflict: Challenges for Societal Stability*, Springer, 2012.

Schon, Justin, and Stephen Nemeth, "Moving into Terrorism: How Climate-Induced Rural-Urban Migration May Increase the Risk of Terrorism," *Terrorism and Political Violence*, Vol. 34, No. 5, June 2022.

Selby, Jan, Omar S. Dahi, Christiane Fröhlich, and Mike Hulme, "Climate Change and the Syrian Civil War Revisited," *Political Geography*, Vol. 60, September 2017.

Selby, Jan, and Clemens Hoffmann, "Beyond Scarcity: Rethinking Water, Climate Change, and Conflict in the Sudans," *Global Environmental Change*, Vol. 29, November 2014.

Shatz, Howard J., Karen M. Sudkamp, Jeffrey Martini, Mohammad Ahmadi, Derek Grossman, and Kotryna Jukneviciute, *Mischief, Malevolence, or Indifference? How Competitors and Adversaries Could Exploit Climate-Related Conflict in the U.S. Central Command Area of Responsibility*, RAND Corporation, RR-A2338-4, 2023.

Siddiqi, Ayesha, "Cyclone Bhola: The Disaster That Re-Made South Asia," *Jamhoor*, October 9, 2021.

Smith, Todd Graham, "Feeding Unrest: Disentangling the Causal Relationship Between Food Price Shocks and Sociopolitical Conflict in Urban Africa," *Journal of Peace Research*, Vol. 51, No. 6, November 2014.

Smith, Paul J., "Climate Change, Mass Migration and the Military Response," *Orbis*, Vol. 51, No. 4, 2007.

Sovacool, Benjamin K., and Walter Götz, "Major Hydropower States, Sustainable Development, and Energy Security: Insights from a Preliminary Cross-Comparative Assessment," *Energy*, Vol. 142, January 2018.

Spillmann, Kurt R., "From Environmental Change to Environmental Conflict," in Kurt R. Spillmann and Günther Bächler, eds., *Environmental Crisis: Regional Conflicts and Ways of Cooperation*, Center for Security Studies, Environment and Conflicts Project, No. 14, September 1995.

Stanley, Jean-Daniel, and Pablo L. Clemente, "Increased Land Subsidence and Sea-Level Rise Are Submerging Egypt's Nile Delta Coastal Margin," *GSA Today*, Vol. 27, No. 5, May 2017.

Start Network, *Disaster Summary Sheet: Bangladesh Tropical Storm/ Cyclone*, April 8, 2018.

Suciu, Peter, "Russia's MiG-29: Why This Old Fighter Just Won't Go Away," *National Interest*, June 2, 2021.

Sudkamp, Karen M., Elisa Yoshiara, Jeffrey Martini, Mohammad Ahmadi, Matthew Kubasak, Alexander Noyes, Alexandra Stark, Zohan Hasan Tariq, Ryan Haberman, and Erik E. Mueller, *Defense Planning Implications of Climate Change for U.S. Central Command*, RAND Corporation, RR-A2338-5, 2023.

Tamnes, Rolf, and Kristine Offerdal, "Introduction," in Rolf Tamnes and Kristine Offerdal, eds., *Geopolitics and Security in the Arctic: Regional Dynamics in a Global World*, Routledge, 2014.

Tekle, Tesfa-Alem, "Ethiopia Vows to Defend Its Airspace from Any Foreign Attack," *Sudan Tribune*, August 15, 2022.

Theisen, Ole Magnus, "Climate Clashes? Weather Variability, Land Pressure, and Organized Violence in Kenya, 1989–2004," *Journal of Peace Research*, Vol. 49, No. 1, January 2012.

Tiihonen, Jari, Pirjo Halonen, Laura Tiihonen, Hannu Kautiainen, Markus Storvik, and James Callaway, "The Association of Ambient Temperature and Violent Crime," *Scientific Reports*, Vol. 7, July 2017.

Toset, Hans Petter Wollebæk, Nils Petter Gleditsch, and Håvard Hegre, "Shared Rivers and Interstate Conflict," *Political Geography*, Vol. 19, No. 8, November 2000.

Toukan, Mark, Stephen Watts, Emily Allendorf, Jeffrey Martini, Karen M. Sudkamp, Nathan Chandler, and Maggie Habib, *Conflict Projections in U.S. Central Command: Incorporating Climate Change*, RAND Corporation, RR-A2338-3, 2023.

Twaij, Ahmed, "Northern Iraq May Be Free, but the South Is Seething," *Foreign Policy*, November 9, 2018.

Urdal, Henrik, "People vs. Malthus: Population Pressure, Environmental Degradation, and Armed Conflict Revisited," *Journal of Peace Research*, Vol. 42, No. 4, July 2005.

van Schendel, Willem, *A History of Bangladesh*, Cambridge University Press, 2009.

vanshilar, "Combat Radius of Western Multirole Fighters," Reddit post, 2016. As of March 9, 2023: https://www.reddit.com/r/F35Lightning/comments/5fv9he/combat_radius_of_western_multirole_fighters/

Vestby, Jonas, "Climate Variability and Individual Motivations for Participating in Political Violence," *Global Environmental Change*, Vol. 56, May 2019.

von Uexkull, Nina, and Halvard Buhaug, "Security Implications of Climate Change: A Decade of Scientific Progress," *Journal of Peace Research*, Vol. 58, No. 1, January 2021.

Wallensteen, Peter, "Food Crops as a Factor in Strategic Policy and Action," in Arthur H. Westing, ed., *Global Resources and International Conflict: Environmental Factors in Strategic Policy and Action*, Oxford University Press, 1986.

Wheeler, Kevin G., Marc Jeuland, Jim W. Hall, Edith Zagona, and Dale Whittington, "Understanding and Managing New Risks on the Nile with the Grand Ethiopian Renaissance Dam," *Nature Communications*, Vol. 11, October 2020.

Wheeler, Kevin, Marc Jeuland, Kenneth Strzepek, Jim Hall, Edith Zagona, Gamal Abdo, Thinus Basson, Don Blackmore, Paul Block, and Dale Whittington, "Comment on 'Egypt's Water Budget Deficit and Suggested Mitigation Policies for the Grand Ethiopian Renaissance Dam Filling Scenarios,'" *Environmental Research Letters*, Vol. 17, No. 8, August 2022.

Williams, Jennifer, "The Violent Protests in Iraq, Explained," *Vox*, September 8, 2018.

Winchester, Jim, "Why the Egyptian Air Force Has Such a Varied Fighter Fleet," Key.Aero, July 27, 2022.

Witsenburg, Karen M., and Wario R. Andano, "Of Rain and Raids: Violent Livestock Raiding in Northern Kenya," *Civil Wars*, Vol. 11, No. 4, December 2009.

World Bank, "Ethiopia's Transformational Approach to Universal Electrification," March 8, 2018.

Yeeles, Adam, "Weathering Unrest: The Ecology of Urban Social Disturbances in Africa and Asia," *Journal of Peace Research*, Vol. 52, No. 2, March 2015.

Yoffe, Shira, Aaron T. Wolf, and Mark Giordano, "Conflict and Cooperation over International Freshwater Resources: Indicators of Basins at Risk," *Journal of the American Water Resources Association*, Vol. 39, No. 5, October 2003.

Young, Oran R., "Whither the Arctic? Conflict or Cooperation in the Circumpolar North," *Polar Record*, Vol. 45, No. 1, January 2009.

Zaheer, Hasan, *The Separation of East Pakistan: The Rise and Realization of Bengali Muslim Nationalism*, Oxford University Press, 1994.

PHOTO CREDITS: Cover: Protesters close Al-Khulani Square, declaring civil disobedience in the October Revolution, Mondalawy/Wikimedia Commons (CC BY-SA 4.0); page viii: Frankincense trees in Salalah, Oman, katiekk2/Adobe Stock.